"A first-class book, full of vigour, ideas and outrage…riveting."
TORONTO STAR

"A clear and perceptive analysis that cuts through the glamour and glitter of the 'brave new world' of communications to show clearly how we can take back our voice."
KINESIS

"Written with uncompromising humanity…an intelligent critique."
EDMONTON JOURNAL

"I doubt there will be a more challenging or valuable book written in Canada this year."
CANADIAN FORUM

"Menzies cogently critiques a discourse on technology that is controlled by a corporatist agenda and offers, in response, a grounded critical discussion informed by storytelling and lived experience."
ANTIGONISH REVIEW

"A compelling and well-researched manifesto for change."
OTTAWA CITIZEN

"An earlier triumverate of Canadians—George Grant, Marshall McLuhan and Harold Innis—warned us about the monsters and hidden dangers in technology. Following in their path, Ursula Franklin, Arthur Kroker and Heather Menzies have continued the vigilance in the new information world.
HALIFAX DAILY NEWS

"A shrewd commentator on the ways in which technology is transforming our lives."
KINGSTON WHIG-STANDARD

Heather Menzies

Stress AND THE CRISIS

OF MODERN LIFE

NO TIME

Douglas & McIntyre

VANCOUVER/TORONTO/BERKELEY

To my precious son Donald Noble Burton,
who has taught me the patient strength of love.

Douglas & McIntyre Ltd.
2323 Quebec Street, Suite 201
Vancouver, British Columbia
Canada V5T 4S7
www.douglas-mcintyre.com

Library and Archives Canada Cataloguing in Publication
Menzies, Heather, 1949–
No time : stress and the crisis of modern life / Heather Menzies.
Includes bibliographical references and index.

ISBN 1-55365-045-X

1. Technology—Social aspects. 2. Stress (Physiology) I. Title.
T14.5.M465 2005   303.48'3   C2005-900706-0

Library of Congress information is available upon request

Editing by Lucy Kenward
Cover and text design by Jessica Sullivan
Cover photograph by Neil Beckerman/Getty Images
Printed and bound in Canada by Friesens
Printed on acid-free paper that is forest friendly (100% post-consumer
recycled paper) and has been processed chlorine free.
Distributed in the U.S. by Publishers Group West

We gratefully acknowledge the financial support of the Canada Council
for the Arts, the British Columbia Arts Council, and the Government of
Canada through the Book Publishing Industry Development Program
(BPIDP) for our publishing activities.

CONTENTS

PREFACE

I BOARD THE TRAIN in good time, find an empty seat and prop my infant son, Donald, in the corner by the window. Then, giving him a big smile and saying "Mummy will be right back," I dash off to make a phone call. The station is crowded and the phone is hard to find. The call takes a while; there are things I have to arrange, right away. Lots of things. Not that I can't handle it; I can, no problem. Some noise intervenes—the train is leaving the station! "Donald! My baby!!" I drop the phone and race onto the platform. The train is already moving. I run toward it, faster and faster, but the train keeps picking up speed. It's moving away, and still I keep running, my chest racked with pain.

Then I wake up. I lie there, shaken. And, as always happens when I have this dream, I tell myself that this would never in fact occur. I'd never lose touch like that, abandoning my own flesh and blood without even realizing I was doing it.

I've been moving toward this book ever since that recurring dream, alert to the sense it evoked in me. I've been asking myself: Who is the baby? What is the nature of the train?

In a sense I've been writing about speeding trains ever since, as well. This is my fourth book probing how society is changing as it entrains itself in instant, globalized communication. And finally I think I've got it—if not the answers, then at least what we need to be concerned about, and why. It's no coincidence, I suspect, that my thinking has become clearer since my son became very sick. The experience stopped me cold and shattered every assumption I'd merrily had about life. It taught me to pay attention, to focus really hard on what was going on and to dwell in it fully, with all the honesty I could muster, if I had any hope of understanding what I could do to help. I have learned to bear witness at a depth I could never before have achieved and, through it, to sense the deeper patterns of what's happening to the world around me.

I write this with a deep sense of identification. I am both the baby abandoned on the train and the woman dashing off to make another career-advancing phone call. What to do will flow from the tension and dialogue between the two.

ACKNOWLEDGEMENTS

HIS BOOK has been a long journey through time and space, with immense support along the way ranging from the intellectual, spiritual and emotional to the financial and the furnishing of facts. Thanking everyone will be impossible, since some support was rendered anonymously, and invariably over the six years this book has taken me, I will have forgotten and failed to make note of some important people.

To begin, I must thank Scott McIntyre, my publisher, who immediately evinced such enthusiasm for the book's ideas that he offered me irresistible contract terms, including the services of the best editor available to help me articulate my ideas and bring them alive—I hope—in vivid, accessible prose. To the extent that I have succeeded, I have to first thank Saeko Usukawa whose intelligence and intellectual acumen drew me out, and whose gentle sensitivity kept me going through times of confusion and self-doubt. I also thank Lucy Kenward whose lucid intelligence and patience with detail took the manuscript through the final editing stage, ironing out the convoluted and the self-indulgent while honouring my particular voice. Finally, Wendy Fitzgibbons added an extra polish to the prose with her deft hand and eye in copy-editing. Thanks to you all, on behalf of the readers, not just myself.

I am almost equally indebted to the many people who helped me in the research and thinking phase of this book. Thanks to David Cayley and Dona Harvey, both for encouraging words and for sharing the contents of their files. For more information and related support, thanks also to Barbara Adam, Usman Ahmad, Robert Babe, Ellen Balka, Jennifer Bayne, Lillian Bayne, Bob Chafe, David Charters, Jacquie Choiniere, Linda Duxbury, David Firman, Bill Fyfe, Bob Glossop, Fern Goldman, Doris Grinspun, Steen Halling, Hanif Karim, Gail Martin, Brian Milani, Len Poirier, John Rager, Andrew Reddick, Vanda Rideout, Wendy Robbins, Ellen Rose, Wendy Scholefield, John Smith, Kit Szanto, Nigel Thrift, Margot Young, plus the inter-library loans staff at Carleton University's MacOdrum Library, Giselle Lacelle and Adam Seddon at the Canadian Policy Research Networks and Bruce Campbell, Ed Finn and the research staff at the Canadian Centre for Policy Alternatives. I extend the same thanks to all the people I quote in the book in conversation with myself, who gave of their time and thoughts so generously.

For help setting up interviews and conversations, I am indebted to Mary Cameron, Raymond Chretien, Joe Courtney, Toni Gilchrist, Elaine Lane, Nicole Mitchell, Tony Pearson, Ginette Lemire Rodger, Hilda Swirsky, Ian Thompson, Dave Tilley, plus many whose anonymity is a necessary tradeoff in a precarious work environment. A special thanks to Sid Schniad both for excellent information and introductions. Thanks also to Jake Marion for help in Greek translation, and to Lisa Greaves for some fecund brainstorming over the book's title. I am particularly indebted to Janice Newson for her assistance and support as a research partner in exploring academics' changing work environment, and to Carol Lane, research assistant extraordinaire.

For useful comments and feedback on early draft material, I am grateful to Sandy Frances Duncan, Madeleine Dion Stout, Nancy Jackson, Farhat Rehman, Denyse Roberts, Bessa Whitmore and Susan Yates. And for support and encouragement throughout,

I want to thank my former publisher Linda McKnight and my friends, including Bessa, Farhat and Sandy, plus Chris Clark, Rita Donovan and Ursula Franklin, who has been both an empathetic friend and a generous mentor to me for more than 20 years.

Last but certainly not least, I want to thank the following granting institutions for the financial support that allowed me to take my time in doing this book. These are: the Canada Council, the Ontario Arts Council and the City of Ottawa's Literary Arts Program.

INTRODUCTION

$\mathcal{T}$HIS BOOK began with a simple observation: At some point in the 1990s, friends and associates no longer had time to get together to just chat—unless it was about work, with a deadline looming. "Did you see that movie/that play? Did you read that book?" I'd ask. "No time," they'd say, heaving great sighs of martyred resignation. No time even for themselves. More sighs. Going to the gym to work out became just one more item to squeeze in.

The people I knew weren't just chronically rushed and overextended, they were scattered, too. "I forget things between the parking lot and the office," a frazzled colleague at Carleton University in Ottawa lamented. "You're always just putting out fires," a friend in franchise services said. "You say, 'Here, try this; I'll get back to you.' But you never do!"

With e-mail, cellphones and the Internet, staying in touch, staying involved, dropping in to check something out, to make new deals, is suddenly so conveniently at one's fingertips. Yet all this contact can quickly be fragmented, becoming mere moments of connection, bits of involvement here, there and everywhere, leaving us with only a vague sense of coherence. And the echo effect, in unanswered e-mail, in phone calls returned as voice mail, in unfinished business and deferred intentions piling up, is creating a miniaturized Tower of Babel in people's heads—a confusion of personal and business priorities, voices and digital bits and pieces. Stress and burnout have reached epidemic proportions, as has depression, with its telltale symptoms: inner disintegration of focus, loss of control and sense of self.

There are reasons enough for all this stress, including two decades of cutbacks, restructuring and doing more with less that have pushed the average work week closer to fifty hours and led what could be a majority of salaried people to put in four hours and more of additional work at home. Plus, life itself seems to have sped up. There are the obvious sources, including the progressive enhancement of travelling speeds, the compression of space and time. Whereas it took months to cross the Atlantic Ocean by sailing ship, weeks to do so by steamship and days by motorized ship, by jet plane the time was reduced to mere hours. Ditto for sending a letter. Now, with the Internet and on-line connectivity, both time and space are said to have been annihilated. The whole wide world is here, *now*. With the letter transformed into digital symbols, communication is *instant*.

The effect of space-time compression is much like being beamed up and down in a *Star Trek* scene. As we leave a voice or text message, or press the "send" button on a querying e-mail, a facsimile version of ourselves moves along a digital network. At any one time we may have many virtual selves in many different places. Of course, simultaneously, others are doing the same thing—dispatching facsimile versions of themselves. The result is that our lives are busier and more distracted because all these various people, or their abstract stand-ins, keep showing up in our e-mail inbox or on our fax machine where they are effectively waiting on hold for us while we're on another call, in a meeting or simply away at lunch. Even though they're not real people, their demands are real enough and they're stacked up in an endless present moment that's synchronized to a vast global network of digital machines that never go out to lunch, never even sigh with ennui and are always at some level "on"—and going full tilt. The

Tower of Babel becomes a welter of real and virtual voices, some attached to who we really are, some absent.

As I listened to my friends and read about all the stress and burnout being documented by the press, what struck me was how people didn't just push the "off" button and slow down. Why is it, I thought, that so many people keep going beyond the first warning signs of headaches, insomnia and lapses in focus and memory? It isn't as though stress is a minor lifestyle complaint, I discovered when I probed its history and found that in World War I it was called shell shock, or "battle stress." I learned that its hallmark is a kind of anaesthetization, even self-forgetfulness. The research suggests that a numbness sets in, an inexorable disconnection from one's self. People lose their inner sense of coherence and equilibrium. Furthermore, this inner disintegration shows up at the level of our body's biochemistry and essential neurological functions, including the "sleeping-brain dialogue" of dreams through which, research shows, we effectively weave yesterday's past into our deep past and wake up not only refreshed but whole again. Notwithstanding this terrible toll, or perhaps because this toll is being exacted, many people drive themselves to the point of collapse, either from chronic fatigue or another breakdown, or to outright death from overwork. The latter has become so prevalent in Japan that it has garnered its own neologism, *Karoshi*.

Many people feel they have no choice but to keep going: they're mortgaged to the hilt, they've maxed out their credit cards and they're putting their kids through school. Plus, well… fast, convenient, high-end living is cool. It's seductive, too: A fast pace and equally virtual way of being in the world are being internalized as an agile, multi-tasking lifestyle. Nor is there much by way of alternatives, with cutbacks in social spending making it harder and harder to qualify for employment insurance or welfare, or to live on the amounts provided. But there's more to the problem of stress and not being able to ease it. For me, it was a fitting coincidence that *Karoshi* was added to the *Oxford English Dictionary* in

1997, when the Internet was declared a mass medium, because the stress that has become so widespread is almost literally spread throughout the networked environment in which so much of daily life takes place. This isn't a stress we're used to. It's not something discrete, like a death in the family that as individuals we can readily isolate and deal with. The stress of space-time compression is not a quantifiable thing at all, but more qualitative. It's integral to the new 24-7 environment, and as more and more of us, or our facsimile selves, are immersed in it we can't help but be affected.

The key for me in connecting stress to our social environment was realizing that today's way of being, both on-line and off, hasn't just accelerated the speed at which people's messages and facsimile selves move through space in time; it's compressed the actual experience of daily existence and saying what we have to say. The speed-up is built into the very medium in which the messages are written, commitments undertaken, performance measured, *and so are its anaesthetizing effects*. They are integral to the trade-off involved in creating a pseudo, facsimile presence here, there and everywhere; for example, in contracted-out project teams or in on-line games and chat rooms. Like it or not, the medium itself exerts a bias toward superficiality and human disconnection even while connection and being in touch, technically speaking, have never been easier. This is because the medium itself packs a bias, or message, if you understand "medium" the way communications scholar and literary critic Marshall McLuhan did, as environment, the material through which we articulate our existence and express ourselves. Knowing this, I realized that the numbing, and all that follows, is implicit in how we represent ourselves in the flickering medium of data on the Net which can move at the speed of light precisely because they are only symbols of ourselves and reality.

Ideas, thoughts, embodied memories and feelings—they're all stripped down to codable essentials, to pure information as symbols in text messages and e-mail, as data fields to be completed or as multiple-choice answers on shopping, gaming or application forms. All these bits of information and data represent reality, but in a way that subtly changes it. As with telegraphic communication, reality is divorced from the rhythms of engaged conversation and the complexities of real life, and that is both its strength and its weakness.

These bits of symbol-information can get around, can get their point across so quickly and efficiently precisely because so much has been left behind in the abstracting, compressing process. They embody, so to speak, space-time compression, making it a feature of our everyday lives. Yet they can numb the senses as a result. After all, there's nothing on which the eyes can dwell, nothing to engage the ear, the nose and the inclination to touch and be touched. When the story being told is just symbols cut off in time and space, there's nothing much to grasp or to become involved with as a whole person relating to a shared reality with others. And if there's no time to re-engage with the larger picture and gain a sense of being fully present, if the next call is waiting, the next meeting is scheduled, the reality in front of you can largely remain just flash and surface—click, on to the next thing.

The problem, I began to realize, isn't just that through cutbacks, downsizing and restructuring most of us are working harder and longer than we were fifteen years ago. We're dealing with bits and pieces of people and situations disconnected from the whole picture. Compounding the anaesthetizing danger, we're dealing with symbols representing those bit-encounters, those fragmented moments of presence and responsibility that come together in the speed-of-light environment of cyberspace (sometimes called hyperspace) to create a reality of their own. As a result there's not just one reality going on, but two at least. There's the living matrix that involves moving as whole human beings through the events of each day, from the wake-up rituals with partners, children or pets and going to work or to a meeting and pausing to say hello to various people along the way, to dealing with various scheduled events. Then there's this other pseudo reality all

5

around us, an invisible virtual or hyper reality composed of symbols standing in for lived, embodied experience yet totally cut away from it and floating free above it (hence the prefix "hyper" to describe it, which means "over" and "above" as well as "excessive"). In fact, this reality can move and change at lightning speed precisely because, like lightning, it is composed of pure energy. It's not composed of matter, but almost its opposite: the cool immateriality of digital os and 1s.

This second reality not only overlaps the one in which we live and talk to each other, make decisions and get things done, it penetrates that space—in the bedroom, in the kitchen, in the car, in our back pocket or our purse—snipping away at the time associated with just being in our bodies or with others. It also infiltrates the in-between spaces and times in our lives where (and when) we could normally catch our breath and regain our perspective. Not only are its demands on our time and our attention constant, this symbol-based version of reality is constantly present too, distracting us with its demands and pulling us in—so much so that we can lose track of ourselves and the connection to the realities the symbols represent.

It takes time to unpack a message. It takes shared time and space, too. It requires not only that people talk but also that they listen to each other, interpreting data and working out what needs to be done. And many people don't have this kind of time any more, or the opportunities of shared time and space in autonomous work teams and institutions. All that's real to them is often only what's in front of them right now, on the screen of the BlackBerry in their hand, the laptop at home or the cellphone in the car: the data sets needing to be updated, the outsourced task or contract to be completed and the performance indicators, maximized regardless of their meaning in the larger scheme of things. Here's where a deeper numbness can set in, through the disconnect from lived reality.

We stay on-line more and more, to cope. We stay in the realm where everything is fast and facile, all simplified symbols allowing us to be present here, there and everywhere as fleet and fleeting facsimiles. We act as though this second reality is real, or real enough, as we take on the now-here, now-there timelines and the tempo of doing and deciding that the simulations make possible. So much is fast and brief, abstracted signs and symbols separated from the often time-demanding complexities they represent that we can lose track of the difference. We can even help foster this forgetting by acting as though we're fully present and by meeting deadlines even if it means largely going through the motions of meaningful participation. For many, this has almost become the only way to get things done. And if the measure and meaning of what's to do is rendered in the same terms, data divorced from the larger lived context, it's easy to forget about the realities on the ground to the point of neglecting them altogether.

Thinking about the broader implications, I began to realize that the symptoms and signs of stress were writ both small and large. I sensed a link between stress as a disease of our times and stress as a more generalized symptom: of social institutions that are losing their integrity and of a society that is losing touch with what's real and what really matters. Grasping this link, I thought, might even help unscramble the crisis of accountability and mean-ing that threatens to paralyze society today.

The key lies not just in the anaesthetizing effect of speed-up and the discontinuities of fragmentation, multi-tasking and asyn-chronous communication. Equally important is how the anaes-thetizing effect built into the prevailing (digital) medium affects us as collectivities with a shared sense of what's real. It's tucked right inside the abstract symbol-language and the point-and-click grammar through which so many institutions conduct their affairs. Without checks and balances, it can threaten the integrity of institutions as organic wholes with shared memories, traditions and values. It can introduce a profound disconnect from a shared social reality.

Some of the most dramatic signals that this alienation might be happening can be found in public services such as education and health care, though this is hardly surprising. The professionals employed in these areas have traditionally defined their work a lot through direct involvement with others, and this can't readily be abstracted and quantified through data, if at all. Yet through a combination of funding cutbacks plus investment in on-line technology, many institutions in these areas have been restructured. The work to be done, how it is managed and how accountability is measured have been redefined, often with data and other virtual forms of reality at its core. From what I've seen, read and heard, the restructuring is profoundly changing these institutions. People can get used to working with e-mail, voice mail and on-line consultations on this and that file or e-contract, yet with each encounter so brief and so superficial that they can't feel fully committed to or responsible for the reality all the bits refer to. Particularly when they are rushed and working in short-staffed, chronically overworked environments, people can also start to act on the pseudo realities on the screen, even identifying with them as more real than the complicated stuff of real life. And so a coroner's jury in 2001 confronted a paradox: While a series of health and social-services professionals posted notes, left and returned phone messages and generally tracked the case of an underweight baby born to a teenage runaway in Toronto, the baby, Jordan Heikamp, starved to death in their care and in a government-approved shelter.

It's become a cliché to say that we live in fictional times. In the news media, supposedly committed to reporting the facts, journalists at the world's most prestigious newspapers get away with egregious fabrications of news stories, not once or twice but routinely, for years! A U.S. presidential administration feeds false and misleading information to the media, which is run as front-page news, then quoted back by White House officials, conjuring consent for a

8

U.S.-led invasion of Iraq. Although such revelations are alarming, to me the more frightening manifestations of this are closer to home and more banal. As the Enron and Arthur Andersen Consulting accounting scandal revealed, data are almost routinely massaged or given a quick spin to direct the public's perception toward EBS (everything but the bad stuff) earnings. And why not, when quarterly profit projections and "fast cycle times" outweigh virtually all other considerations? And when accountability is a closed loop running from expected outcomes to performance indicators and back again. If it's all just fast numbers, all just a shell and bean game, why shouldn't an accountant pressured to meet out-performance numbers change a few assumptions to lower reported expenses? And why shouldn't an otherwise honest teacher simply toggle a factor to adjust the grade point average, or a social worker fill in certain blanks, or not, in monthly reports that *must* be submitted on time? What's a little misrepresentation when the reality being represented is so many times removed from the reality on the ground and the people trying to make some sense of it there? And anyway, so the implicit rationalization goes, everything changes too fast to unscramble the connections even if you wanted to. By tomorrow it probably won't matter.

Accountability has gained importance these days under circumstances of increasingly anonymous units of subcontracted and outsourced work simultaneously under the scrutiny of more and more remote auditors. It's also become more important in public services, in a time of tight money plus righteous expectations of value for taxpayers' dollars and the demonstrable delivery of service to individual citizens. But as Janice Stein argued in her CBC Massey Lecture "The Cult of Efficiency," if accountability itself is not connected to the fundamental human and social values underpinning our social institutions but is left in its own self-referential orbit of technical measures referring to technical values such as efficiency, the whole business can become little more than fiction. A crisis of accountability becomes a deeper crisis of meaning.

Yet who's to stop this or begin to confront its alienating effects? In many of our public institutions, both how we establish meaning and how we verify it are changing. In the past, we talked to each other to clarify our values and priorities. We shared our opinions, challenged our assumptions, asked questions of each other. Now we rely on media, information systems and outside, rule-bound authorities. Our children are growing up without the social skills for engaging directly with others and working things out in groups. Young people at college and university are not getting the opportunities they need to stumble toward expressing themselves and to learn how to think things through in dialogue with mentoring teachers and each other. In an age when time is money, the time for listening and reflection is atrophying. The pace that face-to-face dialogue requires, including pauses for gathering one's thoughts, is becoming too boringly slow. And so dialogue is being replaced too often by data and interpretation by PowerPoint presentations.

I spoke to an academic at York University who said that she is falling more and more silent at meetings these days because she has no time before and between meetings to talk things through with colleagues, to test her analysis and refresh her perspective so that she knows what she wants to say on the current topics. "There's no one to push back on me and say, 'That's good as far as it goes, but have you thought about this?'" Everyone's so busy, and hunkered down behind their computers. She also has little time to attend conferences and keep up with journals. "So even that day-to-day knowledge generation that comes from 'Did you read so-and-so, and this is what they contributed' is lost. There's less and less of that little dialogue."

The civic landscape is falling silent, too. Government has grown remote and ruled by experts, technocrats who deal in jobs, jobs, jobs, and a host of statistical indicators the public is expected to trust as a sign that all is well. Yet when the data are wrong, there's nothing to fall back on except blind distrust and hysteria.

This, to me, is the deeper tragedy that occurred in Walkerton, Ontario, in May 2000, when a lethal E. coli bacteria seeped into the town's water supply, causing half of the town to become ill, killing seven and leaving hundreds of others with lingering disabilities. Yet the townspeople didn't know how to pool their knowledge and take collective action.

Such a combination of hysteria and powerlessness strikes me as a strange kind of civic silence, a noisy silence. It's like a parking lot with car alarms going off left and right while people walk by paying no attention whatsoever, whether plugged into their cellphones and Walkmans or not. It's the silence of a dysfunctional society, one that can plunge into chaos or, just as easily, drift into dictatorship as the best alternative.

From stressed and scattered individuals to stressed and scattered social institutions, from people disconnected from themselves to a society disconnected from a sense of what can be trusted as real and an ability to act on what matters: This is the journey that became this book. I began with only a gut sense that what was happening on the personal level was reflected at a more public and institutional plane. And I proceeded largely on my instincts as well, talking to all kinds of people, devouring whole library shelves of information on the history of space-time compression, the ancillary histories of communication and human expression, and of pre-clock time that was (and still is) embedded in the ebb and flow of life, and in the rhythms of work, of song and dance, of storytelling and conversation.

The result is partly a memoir, in that I've allowed my personal, embodied and sometimes passionate sensibilities to guide what I've taken on and how I've chosen words and metaphors to tell the story of my discoveries. I'd also describe it as speculative nonfiction, in that I not only name the dangers in the midst of opportunity that I sense unfolding in our globalized, post-industrial, postmodern times, I speculate on what these might imply, separately and as a gestalt.

Part I offers the story of stress as it has largely been told so far: how both it and its associated illnesses, workaholism and chronic fatigue syndrome, affect people mentally and physically. However, by placing these new ailments in the context of our globalized hypermedia environment, of which I provide a quick historical introduction in Chapter 1, this section also makes clear that stress is both disease and symptom of a society increasingly out of touch with itself.

Part II picks up on this idea, suggesting that what people take to be real is shifting from what's directly in front of them or discussed in face-to-face dealings to something more remote, more scattered and more symbol-based. An initial chapter speculates about virtual reality and how, as our consciousness is increasingly immersed in facsimile worlds, we could neglect and even abandon the real world.

Part III pursues the implications. It looks first at how children learn to engage in the world and how, as the seeming epidemic of attention deficit disorder suggests, traditional patterns might be shifting. After considering the changes in primary and post-secondary education and how these might be exacerbating rather than mitigating new trends toward more fleeting and superficial forms of social participation, a final chapter brings the themes of reality, attention and engagement together to consider not only a crisis of accountability but of meaning.

Part IV addresses what we can do, as individuals, as institutions and through public policy to redress the situation.

I realize that the crisis of meaning I've detected in the future-present of our world is not particularly new. There are recurring moments in history when things get out of whack, when the pace of change and the new patterns it introduces outstrip our ability to adapt in both senses of the word: adapting ourselves to change, plus adapting it to us in a way that reasserts the primacy of enduring social values. Out of the anguish and distress, out of the accumulating evidence of crisis, a dialogue takes shape. Reforms are suggested, resisted and eventually negotiated. Slowly a new social contract is enunciated, and equilibrium is more or less restored.

Dialogue is critical to renegotiating the social contract, because it puts human experience and, with it, the body back into the picture. It restores data to the context from which they came. As a medium, dialogue also recovers the shared time and space and the pace of considered human engagement. To that end I have written this book as my half of the dialogue, as a recounting of the journey I took, in dialogue with others.

Nearly a century ago, in his classic *In Search of Lost Time,* French novelist Marcel Proust penned a sort of pilgrimage into what he considered to be a lost time of organic unity and coherence in society. He began it with an image of himself dipping a small traditional cake, a madeleine, into his tea, which evoked memories of having done this same thing as a boy when visiting his grandfather. The madeleine was, for Proust, a perfect metaphor for the journey of his long novel, for the shell-shaped cake was a traditional souvenir for pilgrims on their way to a holy shrine in Spain.

Since the journey I'm on is into the maelstrom of early twenty-first-century space-time, I offer the truck as my opening metaphor: a large, long-haul container truck. Join me if you will. Grab a coffee and buckle up.

Part ONE

INDIVIDUALS:

Trashing the Body and the Mind

through Stress and Overwork

One

BUILDING AN
ENVIRONMENT IN MOTION

"We're bringing you the shit you need to survive. We're making the world economy go around. You know how hard it is for us out here?"
BOB WEBB, trucker

"The difference between human beings and their instruments disappeared when saved time was valued more highly than given time."
ARNO BORST, The Ordering of Time

GRAB THE CHROME handlebar on the side of the cab and, reaching for the inside door handle with the other hand, scramble up the steps into Wayne Bates' truck. A quick note in his logbook, then he's ready. He shoves the stick shift into gear and leans his bulk over the hub and spokes of his steering wheel. He checks in the mirror, then reefs on that wheel, clearing his rig through the gate of the freight yard, down the access ramp onto Highway 401 and straight into the fray of Toronto's rush-hour traffic. His eye sweeps the dials on the meters laid out before him, as many it seems as on a flight deck. He switches on the CB radio, adjusting the dial to weed out the static. He opens the case that holds his favourite music—a mix of country, gospel and 1960s tunes. He lights up a cigarette, then pulls off onto the feeder-lane road, and soon he's motoring on by a stretch of stop-and-go congestion. I glance at his profile; he's chatting away about gas prices and so many cents a mile depending on the load. He knows exactly what he's doing, weaving in and out of the traffic, moving up to pass, going down the road. He's king of the road, a modern-day Odysseus riding "the highways of sea water," as Homer described the ancient sea lanes in the *Odyssey*.

The sense of freedom and even of adventure probably hasn't changed much over the centuries, though the power of locomotion certainly has, accelerating from sail to steam to internal combustion and, now, jet engines, electricity and the nanosecond speed of data. Plus, this Homer's Penelope is only a phone call away. Once we've cleared the rush-hour traffic, Wayne grabs his cellphone and calls his wife. He needs a new starter engine and knows a dealer down the road in Mississauga who might have a reconditioned one for sale. Could she check into it and get back to him?

Another difference now is that the entire world seems to be on the move. With sailing ships, which were central to the mercantile phase of global expansion, the goods were luxury items such as gold and spices, plus the grotesque indulgence of slaves. In the days of trains, steamships and industrialization it was farm machinery, lumber, flour and paper, plus the raw materials from which these items were manufactured. Today, with everything from ships and trains to trucks and planes plus the Internet in on the transport act, the goods are almost everything you could possibly imagine, from books to shoes, ankle socks to kiwi fruit, from all over the world. In 1999, 64 million container units moved through the world's fifteen busiest ports, three of these in China, three in the United States. By 2003, between 8 and 10 million containers were being unloaded in New York City alone, though unless their crews had U.S. green cards they were forced to stay on board the freighters as an anti-terrorist "Homeland Security" precaution. Millions more containers move by train across the North American continent and by truck across the border from Monterrey,

Mexico, to Laredo, Texas, while others cross the Canada-U.S. border at Detroit or Sarnia, motoring up and down what's known as the NAFTA Superhighway, named for the North American Free Trade Agreement that opened up commerce to a continental scale. With bar codes and computerized dispatch lists with which to match them, the containers are shunted from one mode of transport to another and hitched behind trucks like Wayne Bates' for final delivery to Wal-mart or Home Depot, to hospitals or office buildings or fast-food franchises such as McDonald's.

The idea of container transportation actually dates back to ancient Rome, when lions were transported in wooden crates from Africa to Rome for sport. Necessity inspired this hands-off closed cargo unit, which was transferred unopened from shore to ship to shore and, finally, to the Roman Coliseum as the animals made their fateful journey to the circus. Two thousand years later, container transportation and communication have come of age as standardized units moving through standardized space and time, as symbols in the air and as stuff on the ground, and society has more or less adapted to the trade-offs involved. Standardization, a seemingly arcane intellectual and technical exercise, played a crucial role in the story.

THE ORIGINS OF STANDARDIZED SPACE

Take a moment to look at your hand, specifically at your thumb. Marvel at how unique it is; flex it, press it against a hard surface and examine its width across the thumbnail. The French word for thumb is *pouce,* which also means "inch." Now consider your foot, or *pied.* This was the second unit for measuring length. The yard, roughly three feet, is derived from the Middle English word *yerde,* a stick used for measuring, which varied locally in length. These words speak to a time when the measure of a thing was inseparable from the context in which the measuring was taking place, including the people involved. It was highly idiosyncratic, contingent on a particular thumb or foot. It gave the units a cultural significance

19

because "they signified or expressed man, the conditions of his life and his work," according to communications scholar Armand Mattelart, who has recounted the role of metric measurement in the rise of France as a modern nation state. On the one hand, these locally grounded measures could be exploited by those with more power in the community, those whose thumb or stick, therefore, carried more heft. On the other hand, these measures had the effect of keeping the people involved in the picture. They erected borders, requiring negotiation and translation between how units were understood, defined and controlled from one region to another.

Revolutionary France changed all that by creating a unified set of standards to facilitate the free circulation of goods on a new, national scale. It did this through the metric system of weights and measures, based on nature at its most abstract. The kilogram standard was based on distilled water at freezing point, measured, by the scientist Antoine-Laurent Lavoisier, in a vacuum. The metre length was based on first measuring the meridian line running from Dunkirk, France, in the north to Barcelona, Spain, in the south, then reckoning it as a fraction (one-forty millionth) of the full meridian line girdling Earth. (In 1927, France redefined the metre as 1,553,164.13 wavelengths of cadmium-red light—foreshadowing the leap of measures into the immaterial realm of data streams, measurable only in such disembodied terms.)

The new metric system of millimetres and centimetres, metres and kilometres, grams, kilograms, litres and so on levelled local and regional borders. It marginalized the particulars of local material existence, made this reality residual. In a sense it took this reality out of the picture altogether, as it substituted a set of abstract symbols for measuring space. The symbols stood in for real things in real places, yet bore no concrete or sensual relation to what they represented. What's a millimetre to you and you to a millimetre? Exactly. Nothing personal.

Having metaphorically disembedded space from embodied place, metrification then projected the notion of measuring space into a new abstract realm where space could be reconstituted as objective symbol-units, for example × units of coal or cloth for transport across × units of distance. This measured space could also be matched to other standardized units of abstract measure, such as money and time, in standardized rules of exchange and trade. It was all rather anonymous, but efficient. It got the job done with a minimum of haggling over details.

THE ORIGINS OF STANDARDIZED TIME

As with the original words to describe space, *tide,* the old English word for time, captures something of time's root meaning, which is life itself and the direct experience of it. It echoes still in William Shakespeare's *Julius Caesar,* through lines like "Thou art the ruins of the noblest man that ever lived in the tide of times" and, more famously, "There is a tide in the affairs of men which, taken at the flood, leads on to fortune…. " The word suggests rhythmic movement: the swelling toward opportune moments, the break that is change and attendant chaos, then the long pulling back as the tide recedes to its nadir. Day by day, year by year, the cycle of tides repeats itself with small variations as the life within it flourishes and falls away.

At its simplest, then, time is life and the events that give it shape, form, identity and meaning. Time is the seed that germinates and pushes a shoot up through the earth toward the sun. Time is the baby who latches onto its mother's breast and nurses. It's the child growing up and, as an adult, growing old, passing on memories, traditions and stories. As time historian Barbara Adam states simply, "We are time," moving to the rhythms of life within and all around us, or not. The smallest indivisible unit of this time, this internally experienced time as life, is neither the minute nor the nanosecond. It is rhythm itself: rhythm in relationships.

Intuitively, we know this and acknowledge it when we talk of bio-rhythms and someone falling into step with others or rejoining a conversation "without missing a beat."

The foundations of human time, to this day, are not found in the tick-tocking of the clock but in the pulse of life flowing between people and within the body. Time is anchored in the beat of the heart and its systolic (compression) and diastolic (relaxation) phases. "The drum," Algonquin Cree elder Jacob Wawatie told me once, "is an extension of the heart." Long-short, long-short; the basic iambic beat, which has long been the heartbeat of culture, is mimicked in the beating of a drum marking time for dancing or the telling of traditional stories.

For ages, societies lived immersed in this time consciousness. The mediaeval church built on it, with Christmas known as Yule-tide and Easter as Eastertide, and various events like lambing, de-horning cattle and pasturing animals timed around feast and holy days. These annual cycles not only ordered life, they bound people together (for better and for worse) as communities and gave meaning to their shared life on the land. Then the clock came along, first to discipline daily life in monasteries and later to regulate business and commerce in the cities. It uprooted time from the rhythms of tides and seasonal cycles. It separated time from life and turned it into an object with its own value and meaning.

This, in a nutshell, is why urban geographer Lewis Mumford calls the clock the most important invention of the modern industrial age, more important even than the steam engine or the printing press. In his view, the clock "disassociated time from human events and helped create the belief in an independent world of mathematically measurable sequences: the special world of science." Much as metric and other abstract units of measure revolutionized the meaning of space, the clock changed not only the measuring and marking of time, it revolutionized the meaning of time. The core invention, then, was both silent and essentially invisible: a whole new conception of time as symbol-units, an accumulation of abstractions called seconds and minutes, which the clock's pendulum and interlocking gears tick-tocked into being.

At first, this time shift didn't affect people a great deal, at least while clock time stayed on the wall. However, as clock time began to be built into the infrastructures of life, in the assembly line for example, its effects were more personally and acutely felt. Now it pulses all around us, and in the pager on our hip. Clock time is altogether separate from the body and the actual experience of life; yet, Mumford says, it has become the new "medium of existence," the way in which much of life is experienced. Disconnected from the rhythms and events of real life, clock time is also infinitely divisible and compressible into shorter and shorter "timelines."

The standardization of time didn't just happen as a deterministic consequence of invention. Rather, being on time—standardized clock time, that is—took hold because it fit with the general constellation of developments that came to be known as modernity. These ranged from the ideas, techniques and technologies associated with modern science to the ideals of efficiency and rationality. These, in turn, harmonized with and jelled into a philosophy of progress as expansion, speed and material wealth, not just for nation states but for nations defined as aggregates of individuals free to maximize their upward mobility year after year. Physicist Isaac Newton's notions of time, as both separate from space and a measure of motion or duration, lent legitimacy and even privileged authority to the clock. Similarly, the clock helped to advance new sciences such as economics, as it offered a medium for managing abstract laws like supply and demand through production and delivery plans.

EARLY UNDERSTANDINGS OF SPACE-TIME COMPRESSION

In *The Wealth of Nations*, moral philosopher Adam Smith noted that "industry of every kind naturally begins to subdivide and improve itself" when it's located near a waterway where goods can be transported much more quickly and efficiently than on land. Smith is considered a founding father of modern economics for the ideas he expressed in that text, including scientific management and the division of labour. With a view to exploiting the advantages of fast, cheap transportation, work was uprooted from being an integral part of a local, or even an individual's, unique way of doing things. It was broken down into fragments and reconceived as a series of standard actions to be completed within a certain time. The goal was to ensure a standard product and a steadily improving rate of production at a cost that would allow the goods to be marketed farther and farther afield at locally competitive prices, thanks to that fast, cheap transportation. The idea was to compress space (and, later, time) to the point that it was virtually annihilated as a barrier to doing business anywhere at any time.

For well over 100 years, such efficiency measures were pretty well confined to the factory, though they were much refined there, particularly through American innovations such as Henry Ford's auto-assembly line and engineer Frederick Taylor's famous time-motion studies for reducing steps and inefficient motions. Whatever the product being made, the process was honed into a series of routine procedures that were standardized not only into prescribed task-units but into prescribed units of task-time as well. Personal interpretation, or embellishments identifying the workshop or individual craftsperson (called "signatures" in the craft tradition), were streamlined to a minimum, along with any discretionary time required to indulge them. Goods became uniform, interchangeable and anonymous under brand-name labels. With the rise of bureaucracy during the Second World War, these efficiencies were introduced to the realm of services, including public services. In the post-war culture of cars, convenience and mass consumption, the McDonald brothers extended the efficiencies into retail services and the activities around consumption when they pioneered the concept of standardized, assembly-line-like production of fast food. Their Speedee Service System became the backbone of their soon-to-be-famous golden arches.

Of course, none of these changes could have happened, at least not in the same way, without a corresponding standardization of communication and expression, one result of which is that an iconic set of golden arches depicting one letter of the Roman alphabet means cheap food and fast, efficient roadside service to millions of people around the world. I'll get to this later. As for this somehow representing progress, that's merely part of the story.

OUTSIDE THE WINDOW of Wayne Bates' truck, it's twilight. The rush-hour car traffic has faded away, leaving the road to us. We're part of a steady stream of trucks heading east, facing what seems an equally steady flow heading west, two streams of us going full tilt in opposite directions along the highway. We're part of a world in motion, a world in which truckers are expected to meet tighter and tighter delivery schedules so that the parts or products they're hauling arrive just in time for the next shift in a factory or the latest sale in a store, regardless of what's happening in the here and now. Spaces as individual places don't matter as much as the flow of transportation and communication between them.

It's amazing, too, how the physical environment has adapted to this flow, with pit stops along the highway where people click credit-card-like "speed passes" to shortcut their way through a gas-up; historical markers and related picnic sites seem quaint relics from a former way of life. Beyond the multiple lanes and access ramps to the highway, I see the familiar profiles of suburban malls, freight terminals and the occasional factory. The back end of each is identical, its loading-bay indentations just like the back end of a Lego block for quick-click interconnections.

As the last daylight fades from the landscape, I think about how adapting to this environment causes so many things and places to

25

lose what philosopher and literary critic Walter Benjamin referred to as their "aura," their unique "presence in time and space." It used to be that kiwi fruit meant New Zealand and tartan meant Scotland, where craft weavers honed generations of apprenticed knowledge and skill to produce vegetable-dyed fabric woven from the wool of local sheep. There was a time element to the old craft traditions, too; production unfolded at a measured pace to give quality craftmanship its due and allow produce to ripen with the seasons. Now kiwi fruit can come from Chile and tartan is more cheaply made in China, with the skill and knowledge captured ("black-boxed") in computer subroutines and all the inputs assembled through outsourced bidding matched to on-demand orders— that is, contracted out to sources of labour and supply outside the firm and even the country, according to orders relayed on-line from stores or individuals around the world for forty-eight-hour delivery.

When I was hanging around the freight terminal, one of hundreds on the northwest outskirts of Toronto, waiting to catch my book-researching ride with Wayne Bates, I heard a host of accents and saw a range of dress styles among the people ferrying standardized pallets of merchandise into the standardized container-truck units. There was an aura inside that look-alike terminal. Despite the repetitiveness of loading endless lists of goods into designated container-units in time to meet their departure schedules, there was a definite sense of personality and even culture. Yet none of this permeated the plastic bubble packs around all the products these workers were loading. In that sense, they were just as anonymous and just as invisible as the people who produced the merchandise.

The phone rings. It's Wayne's wife reporting on the part he needs for his truck. The numbers match the specifications for what he's looking for, and he can get the piece installed tomorrow afternoon. Great. Wayne clicks to end the phone call, then throttles his engine and signals his intention to pass.

INTERNATIONAL STANDARDS AND
THE DREAM OF GLOBAL SYNCHRONIZATION

The journey Wayne Bates makes every night in his container truck epitomizes how society's constellation of structures and ideas, of modes of doing and being and even of defining what's real and relevant, is changing. A transformation is occurring, not just in world view but in how the world is organized, ordered and run. The space of places—such as nation states and distinct national communities—is being eclipsed, even replaced, some argue, by a "space of flows," populated by new entities such as transnational corporations that transact their business and identify themselves through flows of data, logos and mission statements regardless of place and people's traditional ways of doing things.

Representing space and time as abstract symbols was a radical enough innovation; with the space of flows, however, those symbols become real in their own right. They become what people focus on, especially as they come together as a self-referential sphere of interconnecting symbol-units organized and coordinated by ideas expressed as production orders, supply contracts and investment plans, plus just-in-time delivery schedules and software programs not only coordinating and monitoring it all but quantifying the results. Global digital networks—a combination of corporate information, investment, management, advertising and distribution systems available over the Internet in an easy-to-use graphic format—are the linchpin of this space of flows and the reason that it seems to be everywhere, all the time now, trumping everything else. It's a dynamic sphere of action operating on its own transnational scale, its own interconnectable scope and its own lightning speed; in short, it's a new world order.

A host of innovations, ideas and developments coalesced to produce this immense turn of events, or rather this turn in how events would be managed, recorded and evaluated, within the dynamic space of flows. International standardization has been one enabling factor, especially in how the flow of standardized units through space and time might be synchronized around the world. The introduction of global standard time in 1884 was one of the most significant early initiatives. By itself, this project brought together many vested interests and the institutions associated with them. But equally it was the crowning achievement of one person, Canadian railway engineer and surveyor Sir Sandford Fleming. It's telling that Fleming came to the question of time after a long and successful career dedicated to maximizing flows through space, specifically masterminding a rail infrastructure for moving people and freight across whole countries and continents. In the early days of railway travel, trains were crashing at a rate of almost one a day. They were hurtling along without accurate information about how they would intersect in time, because local times varied from one point along the railway system to another. The local times made sense locally, but for train transport to function efficiently something had to be done to integrate scheduling across the country so that trains could move along a grid of coordinated time as well as space. The trick, Fleming became convinced, was to synchronize the telling of time from one place in space to another.

From the outset, Fleming thought of time in the same way that he had learned to think of space: as abstract numbers for distance and elevation laid out in surveys and rail-line construction plans. He thought of it as a "kind of hyper time," Clark Blaise wrote in an elegant biography of the man. Fleming saw these abstractions as building blocks in a clockwork universe that society could build and run. He himself used the terms "terrestrial time" and "cosmic time," arguing in one paper he gave: "We are now obliged to take a comprehensive view of the entire globe in considering the question of time reckoning." He was among the first to both think and plan in such boldly global terms. His idea was to take the tool of standardized-unit clock time and map it across the entire geographic circumference of the world. Although getting others to support this plan wasn't easy, Fleming figured that the telegraph's instantaneous communication would allow the precise time reading at Greenwich to be sent to "standard keepers of time" in every country to achieve "perfect synchronism," at least in the idealized realm of symbols on clocks and various schedules.

The dream of perfect synchronism had its corresponding champions in the realm of space, where they helped to internationalize standards for weighing, measuring and grading goods to facilitate trade and commerce. During World Wars I and II, the scope of this standard-making expanded radically to include standards for coordinating among the Allies the production of everything from coal and steel to warplanes and ships. After World War II, the International Organization for Standardization (known by the acronym ISO) was formed to facilitate the international coordination and unification of industrial standards and, by doing so, to increase productivity and international trade. It was one of an increasing number of organizations through which regulation by international technical standards has replaced regulation by national policy, all expanding the realm of a global space of flows. During the multinational phase, companies used international specifications in design and production to ensure a certain standard in branch plants around the world, so one source could replace another, and parts or products could be rendered interchangeable.

In the 1970s and '80s, organizations like the Organization for Economic Co-operation and Development, a think-tank on social and economic policies, created international standards for electronic data exchanges that, coupled with deregulation in the banking, currency and investment fields, helped bring on the transnational phase of this space's expansion in developed and developing countries alike. Instead of establishing branch plants in different countries to produce a line of products for sale in those countries over a period of time, transnational corporations concentrated more on the financial-management end of business and issued renewable contracts to a global list of possible suppliers. Today, in hyperconnected, high-bandwidth New York City, nearly

29

$2 trillion changes hands electronically every day. The Bank of America turns over $60 billion in foreign-currency trades in a single day.

Some of the most important standard-setting and re-regulation through international standards and global bureaucracies has occurred at the juncture of computers and communication technology, the juncture of time and space. Standards permitted computers to be linked to create the first information systems, and more standards allowed these to be expanded first into local, large-area networks and eventually the global network associated with the Internet in the 1980s and early '90s. The World Wide Web Consortium was established in the early 1990s just to coordinate standardization on the Web, which with its user-friendly icons and hyperlink connections represented a huge advance in global standards all its own. As one of its projects, an international standard for collaborative Web-linked manufacturing has been developed, featuring an "extensible markup language" (XML) that effectively containerizes instructions for machine production much as "hypertext markup language" (HTML) packages an array of personal and cultural information for instant transmission, location and retrieval across the Internet. Both these examples are not just machine languages, they're a connecting grammar that allows information to be moved across the Net, crossing borders and overcoming language barriers without missing a beat.

Increasingly it's becoming common for factory equipment to have a built-in (standardized) Net connection so that incoming operating instructions can slip not just through the corporate portal but straight into the memory of the machines on the floor. The added advantage is that machines can be monitored from a remote head office just as easily as from an office directly overlooking the shop floor. This, in turn, allows companies to use the entire Internet as an extension of their corporate-management networks to contract out work to sources that provide cheap labour and inexpensive materials—in other words, to "outsource" to factories and firms around the world. Equally, outsourcing allows them to take on new projects, to invest short-term or long-term here, there and everywhere. Thus companies can be truly global, with a standard face and interface helping to make it all possible. Perky, familiar words like "agile" and "flexible" were used to describe this new arrangement, suggesting that both companies and the people who worked for them could change what they were doing, as well as how, when and where they were doing it, with the easy adaptability of a seasoned athlete ready and willing to take on any challenge.

Each step in the expansion of markets and, with this, the standardization, unitization and speed-up of production and distribution, has entailed successive degrees of separation between the people making the goods and the people buying them. Goods and eventually services, too, have necessarily become anonymous so that they can be readily interchangeable, moving according to supply-and-demand fluctuations in local and distant market prices. The idea is to incorporate more and more activities (and people) into the transnational zone of flows and have them all flow quickly and efficiently as standardized units in space and time. From a business point of view, people and production facilities must be readily replaceable with minimal retraining or retooling costs. Then, too, jobs can become equally interchangeable with little or no adjustment, angst or long-term commitment.

Many parties have a vested interest in embracing this trade-off, including the solution it offers to the dilemma of trust. In place of a local product's or person's reputation or character as the measure of accountability, a new set of standards and related abstractions associated with trusted global brands and related logos now stand in. In addition to government grades that measure the quality of products, production itself is subject to standards: everything from operating procedures and ISO-approved business practices to standardized credentials and related skill and course requirements and, finally, to standardized ways to measure and evaluate work, called performance "metrics." Increasingly too, companies are

31

subject to regular ISO audits to ensure that workers are following every step of the approved operating procedures and following the prescribed methods for documenting what they're doing, in every batch and on every shift. Goods or services, it makes no difference. They're all deliverables to be produced and dispatched according to uniform, symbol-based routines.

In the new globalized economy and the hyperworld view that accompanies it, what's important is that the symbols on the various printouts match the symbols defining the best-practices protocol the company involved must follow to retain its standing as an eligible supplier or franchise holder of whatever the deliverable might happen to be. Curiously, this detailed level of auditing is used not just in the manufacture of big equipment, where any error might be dangerous, but in the cleaning of hotel rooms and even the delivery of service in fast-food restaurants and coffee-shop chains. One rather bizarre result is that "quality" now tends to mean "performance to specifications," even where performance means serving people, one on one and face to face. At some fast-food outlets, quality-control supervisors drop by as "mystery shoppers" and rate employees not according to how they rise above the company standard and provide extra-special, extra-personal service, but according to how exactly they follow the standard script on what they're to say and do within the prescribed seconds or minutes. What counts is matching performance to the standards associated with the corporate logo and image, everywhere and every time.

Just as international bureaucracies like ISO and the International Telecommunications Union synchronize the flow of standardized goods and services technically, agreements like NAFTA and the General Agreement on Trade in Services (GATS) facilitate these flows politically, and trade organizations like the World Trade Organization enforce them. All these bodies work to take down any obstacles ("non-tariff barriers") created by national and local policies that would prevent goods and services from being interchangeable as "like" units, not just from a technical but a public policy point of view. For example, GATS negotiators are trying to redefine services such as education as standardized deliverables. By doing so, providing such services could reasonably be open to bids from any would-be investor-supplier of learning, using whatever "modes of supply" are most efficient and competitive, from face-to-face classroom dialogue to global multimedia service centres where learning modules are packaged like any other container unit Wayne Bates hauls down the road at night.

IT'S PAST 10 p.m. now. Wayne and I are somewhere along the 401, though it's hard to tell exactly where. The landscape has been winnowed down to recursive bursts of light signalling McDonald's, Wendy's, Tim Hortons, Esso, Petro-Canada and other look-alike links in the chain of fast-food and fill-up stations. I watch them go by, and think: It's almost immaterial what goes into the latest gadget. It hardly matters whose fingers are on the cash register, whose happy face is offering you "air miles," whose hands are flipping the burgers or steering the truck going down the road delivering all the latest stuff, just in time and right away. Everyone has a name tag but they might as well be nameless. All that matters is meeting the specifications, following the script or quality-control procedures and making the target in space and time. All that matters is the flow of deliverables and, for more and more people in more and more parts of the world, it seems, simply being part of its 24-7 connectivity and keeping up to speed.

I sit back, watching it all go by. My eyes burn from all the blazing headlights; my ears ring from the roar of speed and motion. Wayne flicks his high beams on and off, greeting the driver of another truck hurtling by, and points out the running lights along its cab. It's like a signature, he says. 33

"I see the lights and I get on the radio. 'Hey, Gerry, how you doin'?' Or it's 'Geez, Wayne, you're gonna run into snow down the road.' Or maybe 'There's a bear sitting at Mile such-and-such.' You know. And that's why you see so many trucks with so many different kinds of lights."

I've been riding with Wayne for a few hours now. We even had "breakfast" together at 5 p.m. at the little greasy spoon outside the terminal gates, Greek travel posters all over the wall, before Wayne started his shift. He's been showing me all the features of his customized "Black Hawk" truck cab, including its air-cushioned seats with adjustable arms, back and height, plus the miniature fridge, TV and microwave oven in the sleeping-bunk unit behind us. He's almost paid off the $123,000 rig, he's told me proudly, right down to the few extra dollars he threw into his own set of running lights. So I chance a quip.

"Oh, I thought customized lights were just ego and personal affectation."

He laughs. "Well, it is. There's a lot of that, too. You know—I got this, I got that. A lot of people have too many. A lot don't have enough. Some look gaudy; some look real sharp. It's all personal taste."

He checks the time on the dash. We're right on schedule for a doughnut and coffee at a favourite truck stop. He pulls in and finds a spot in the acreage provided for parking.

"We'll probably meet up with about six people here. Regulars. One fella drives for UPS. Another fella drives for Dixie Transport out of Napanee. It's just the guys you pass every night. Or you see them in the coffee shop. So you give them a nod or a 'how ya doin'?' because you recognize them. So you get talkin' to them on the CB, and next thing you know you're friends."

On this frigid mid-week winter's night, the football field–size lot is nearly full. Drivers on a long-haul run, maybe through Detroit to Kansas or south to the Mexican border, have pulled over for the night. They've had a shower in the facilities provided, perhaps washed some clothes in the laundromat, bought takeout food at the store or sat with friends over a meal in the coffee shop. Then they've bunked down in their cabs, blackout curtains pulled over

34

the windshield, yellow parking lights glowing like cats' eyes, engines idling in a uniform drone. The lot reminds me of a beehive on a hot summer's night, the worker bees fanning their wings in unison to keep the beeswax cool.

In the coffee shop, business will gradually subside after midnight, achieving something close to dormancy at about three o'clock, when the all-night waitresses gather two or even three empty coffee urns in each hand, take them to the back and give them a good scrub. Coffee keeps these truckers going on their ten-, twelve- and sometimes fourteen-hour runs and sometimes eighty-four-hour weeks. Coffee and new pep-you-up soft drinks like "Surge" and "Jolt," and sometimes uppers and other banned substances.

"Everything's just-in-time; there's a lot more pressure now," Wayne tells me as we doctor our coffee with sugar and cream. "Everything's done by appointment." That is, an appointment to make delivery at some company's loading bay. "It used to be, you told people they'd get the goods when you got there, when you made it through the blizzard, the rainstorm, the traffic. If they gave you a hard time, tough. Now, they tell you, 'tough.' A lot of places, if you're half an hour late, they'll just refuse it, 'Make another appointment,' they say. So then you gotta take your trailer back to the depot. And it might be the following day before you get another appointment, so then you're tied up.

"And you can see it, too," he continues, reaching for a napkin, "because they're running tight themselves. If somebody's late, it throws everybody off." He polishes the table where he's spilled some coffee, then looks up. "Like, let's say your appointment is for ten o'clock and my appointment's for eleven. You don't show up 'til quarter to eleven, and it takes let's say roughly an hour to unload. Well, that'll back me up an hour, and you don't know what I've got to do. I might have a pickup to do at 12:30 or one o'clock somewhere else."

Wayne's trucker buddies have joined us, and one of them mentions being fined for "time theft." In the auto industry, these guys tell me, companies can charge $60,000 *per minute* in fines for any production time lost because a line is down waiting for parts to be delivered. A lot of guys, with payments to make on their trucks or their houses or for their kids, take on extra loads auctioned off on the (computerized) board, competing against each other in effect and agreeing to deliver a load from A to B in the impossibly short "average optimal" trip time posted.

Wayne knows at least one man who dropped dead from a heart attack. "He just kept pushin' himself. 'Make that window!' Everything's based on mileage. If it's × amount of miles, it should take you × amount of time, no matter what the weather, what the road conditions, what the traffic on the road."

"Sometimes I feel like a spider on acid," says the trucker who mentioned time theft. "Everything's time-sensitive these days. They need everything just in time."

"We're always chasing time, no matter what we do," a third trucker adds.

They pay for it in more ways than just their personal health and well-being. In one of the more gruesome stories of road rage, a car driver shot and killed a trucker, presumably angry at being slowed down or cut off. More routinely, marriages fail so often that, one guy tells me, there's a betting pool at the trucking firm he works for, with odds on who will be next.

"This business is very hard on marriages," one man tells me while the others nod in agreement. "You're up at 3 a.m. and gone."

Someone checks his watch, and suddenly they're up and at it, their thermal mugs topped up, their cellphones back in their pockets. I follow Wayne outside, past the rows and rows of look-alike container trucks, each one pumping a plume of global-warming carbon dioxide plus respiratory toxins into the air, each one its own miniaturized world, each sporting its own distinctive line of running lights around the cab, its own bit of attitude and identity. So this is progress? I ask myself as I clamber back inside the cab of Wayne Bates' rig and fasten my seat belt.

STANDARDIZED SPEECH AND THE HYPERMEDIA ENVIRONMENT

There's another thread in the story of standardization and space-time compression that I've ignored so far: the standardization of how people express themselves, the compression of how we communicate and achieve consensus on what's real, and what gets left behind or at least marginalized in the process.

To understand this, I need to go back to the time that human communication began to be standardized. When the ancient Chinese and Sumerians invented the first alphabets, they created both the written word and a new abstract medium for communicating: that is, dissociated from the body. It was divorced from the subtle play of body language and distanced too from the person being addressed, whose verbal and non-verbal response might influence what was being communicated. (Admittedly though, writing by hand remained quite personal and idiosyncratic.) Mechanical type pushed human expression more completely into the realm of clocked time and metered space where, like those, it could be further standardized and compressed. Language was rationalized across space, much as time was standardized nationally and internationally. Local vernacular was homogenized into national languages, with standard definitions and, eventually, uniform spellings set down in official printed texts. With mechanical print, precisely shaped letters and punctuation marks were assembled quickly and efficiently into words and sentences. With mechanical presses these could be produced, packaged and distributed widely, as mass-circulation newspapers, popular novels or classroom texts.

What resulted was a trade-off of spontaneity and nuance for predictable meanings, conveyed expeditiously. A certain superficiality resulted, a clipping of the sensitivity to oneself and others that can accompany face-to-face dialogue. The consequences became more noticeable later, when standardization and compression in communication had reached the point where boilerplate and PowerPoint presentations, e-documents and data sets, sound and sight bites and other Lego-like modular language had replaced the time and pace of storytelling, conversation and interviews. But by then the time not only to express oneself but also in turn to listen, to unpack nuanced meanings and interpret these in light of deep memory associations, had been compressed out of a day deluged with incoming data.

The printed word gained authority over the spoken word in part because of the institutions associated with its use, plus the money and power they packed behind their printing presses, and in part too because it resonated with the mindset of modernity. The exact layout of the texts was almost an exemplar of modern science, which championed knowing through objectification, standard classification and abstractions. Subtly, too, as communication became less bound with face-to-face conversation and the giving and receiving of communion it came to be less understood as something intimate, fiduciary and integral to social bonding. Through the twentieth century especially, as the printed word gained prominence, communication came to be understood as the transmission of information as efficiently as possible and with the least "noise" or interference. The centrality of the voice and its integrity subtly gave way to data moving along a transmission line, be it telegraph, telephone or the production line of a newspaper.

Today the transmission line has gone digital, with startling results. Being digital, it is both speed-of-light fast and totally fluid. So it can take any module of information from any medium of communication, from print to video to voice, translate it into its standardized 0s and 1s and draw it into its sphere of transmission, its media space of flows. Moreover, it has merged with everything else in the larger space of flows. It is both a medium of expression and communication and a medium of production, distribution and consumption. It's all one and the same: all data. The data themselves may not be able to "speak" to each other, but the grammar—the point-and-click techniques, the standardized containerizing routines—is the same, and so is the accent on speed. The result is synergy: interchangeable data sets that can move across multiple platforms and operate in a range of contexts. Subscriber orders are transmitted together with cellphone text messages, click-and-send images flow alongside buy-and-sell orders on the electronic stock exchange, while work orders are contracted out to a global or local workforce and deliverables are relayed back on-line or off. They're all standardized, container-like units. It doesn't matter what they are or where they're going either, the transportation and communication costs being about the same.

No wonder theorists talk about a tipping, even a shifting, of paradigms and world views. The space of flows is all around us now, running through the cellphones and hand-helds in our pockets and the pager on our belt, through our cars and our homes, and through almost every institution. It's taken on the aura of an environment in its own right, with more and more activities and parts of institutions plugged into it, on its terms. It's collapsing the difference between inside and outside national borders and, equally, the difference between the public and private individual. This means that as long as you're in that space, your movements are simultaneously being read and tracked. Data-mining firms routinely sort individuals into market modules, or "segments defined in terms of price sensitivity, coupon use, brand loyalty, television use and other characteristics of interest to consumer product marketers." Your identity is both a customer profile to which various goods and services are marketed with "it's really you" enticements, plus what you say and do in whatever space and time you have in which to truly do your own thing and speak your own mind.

Another aspect of the grammar of this new language is that its standardized, interchangeable abstractions can be assembled and deployed more quickly and in a wider variety of ways than ever before. For example, corporate brand names and logos can be embossed onto practically any product—from bandages to baseball caps—being produced anywhere, and they can also be downloaded onto business cards and letterhead as new associates join the team and take on the corporate mission statement. Equally important, the language flows in and out of traditionally separate worlds— from business to health care and back again—with merely a click of a mouse. It's all taking care of business, working the data, playing the numbers game. Expressing yourself means using the medium as much as it does speaking for yourself and in conversation with others; it means using your cellphone to leave a message for a new friend, upgrading and downloading to assemble a look or a resumé with brand-name university or consulting-firm affiliations and with cutting-edge skill sets. It means cutting and pasting to generate a report or a paper as much as it does articulating something new and original from your own considered thought or dialogue with colleagues, teachers and mentors. That's the main message of participating in this medium: it's all about making and maintaining connections, being in the lead and on the go in the digital zone of flows.

As Marshall McLuhan wrote explaining his "medium is the message" aphorism: "The 'message' of any medium or technology is the change of scale or pace or pattern that it introduces into human affairs." In short, he argued that the deeper message of any medium lies within the infrastructure, the operating systems, the symbols and grammar of the medium itself, not what is written on the page or the screen. He maintained that media function in the same way that environments do. They condition us, producing "psychic and social consequences," notably in three distinct ways: through scale (global versus local), pace (instant, lightning speed versus face-to-face rhythms and continuities) and pattern (asynchronous hypermedia and multi-tasking versus organic wholes).

Today, the space of flows has matured into not just a multimedia but a hypermedia environment, in the sense suggested by terms such as "hyperspace" coined in science fiction. That is, when multiple individual lines of media and communication merge, they produce a gestalt effect that is over and above what they could achieve before. It's a space-time environment in which symbol-units representing space, time and human expression connect and interact with each other in their own self-referential orbit. Furthermore, many of these interactions are controlled from within that space by computer software and governed by the Net's built-in efficiency standards of global scale, recombinant scope and speed.

The space of flows is the operating environment of globalized business, and individuals and institutions alike are adapting themselves to it—because they feel they have no other choice. The supremacy of this environment and its bottom-line logic are rarely challenged. Corporations are creating more interfirm alliances to tap the global economies of scale, scope and speed, and they're outsourcing more and more work in order to remain competitive. Institutions faced with ongoing cutbacks are tapping the efficiencies of centralized, information-based management and accountability systems plus the shortcuts of e-conferencing, e-mail, voice mail and text-messaging on cellphones and personal digital assistants (PDAs). For people everywhere, adapting to this hypermedia environment means adapting to ongoing change, to a new system for record-keeping in the hospital, for accounting in the store, for filing and referencing systems and keeping track of information in the office, all with opaque codes that threaten to leave you behind if you can't keep up. It also means adapting psychologically to the trade-offs of fragmented and often symbol-based communication, which precludes the slowed-down rhythms of face-to-face conversation.

In this hypermedia zone of flows, everything is fast and fleeting. Everything is fungible, too: mixing and mashing music is the new trend in culture; cutting and pasting the new trend in production, downloading the new trend in identity. Living in the zone means being a fast subject, being agile and flexible and always ready for change. It means fast positions and opinions, fast relationships and commitments. And it means fast fashions, the new trend in dressing in which companies grab patterns for the latest look on a fashion-show runway, e-mail them to suppliers in China

41

or Romania or wherever they can find standard-meeting equipment and credentialed workers at the least cost and turnaround time, and within days have the new look shipped to clothing stores in London or Toronto, where they're sold at irresistible knockdown prices—such that it's cheaper to throw it out and buy something new next week than it is to take it to the dry cleaner.

THE DANGERS AND COSTS OF LIFE ON THE GO

The adjective "hypermedia" has become a verb. Between its lightning speed and its quicksilver symbols, the zone has become dynamic. It's a medium-environment the message of which is not just a fast "massage," as McLuhan once said in a play on his popular aphorism, but, I think, also an anaesthetizing one. Without the right checks and balances, this living space can hypermediate us into a state of stress, burnout and self-forgetfulness. It can take us away from the realities in front of us, including the micro-environment of our own bodies with their primal needs for rest and relaxation. It's not just the relentless speed of it all, nor being scattered across a bunch of multi-tasking fragments, it's the fact that we're engaged in a realm of pure representation—ready-made icons and modules of standardized symbols that, being symbols only, don't permit any deeper level of involvement. The space of flows can begin to flow right through us, hardly touching our bodies, our memories or the inner core of who we are. It can hypermediate us right past the symptoms of stress and burnout and through the door of numbed self-forgetfulness into a whole new realm of reality (virtual reality) and, with it, a whole new way of being that is all symbol and no real substance, all fast-forward momentum with no pause for rest and reflection.

The promise of virtual connections, of getting around and making things happen in this hyperspace-time, is indisputably exciting. But it's easy to get carried away and to lose the equilibrium we need to be healthy and to know who we are—especially for those of us steeped in the world view where progress is speed, (upward) mobility and convenience. It's all there in today's "friction-free" consumer capitalism: click to make contact, click to make a deal, click to contract out the work, click to arrange the necessary financing, benefits, insurance, royalty payments on any software or logo used in the collaboration, and click to check on its progress. All of these tasks we do indirectly, through computers, hand-helds, PDAs, cellphones and, behind the screens of these, e-mail, file management, electronic documents exchange and e-commerce. All of these actions we effect in the guise of facsimile- or pseudo-persons: log-on corporate-player or shopping-points club identities, or various aliases for sex, gaming, gambling and other dealings. These are simultaneously real and yet strangely anonymous, removed from any means of sensing genuine presence, commitment and responsibility for something beyond the symbol-moment.

This hyperworld is a parallel universe that exists on-line, and only there. Yet it has taken on the aura, and the authority, of reality as it spins into its own orbit, increasingly cut off from real life in all its challenging complexity. It's a realm where matter doesn't matter, because matter with all its sticky friction of body against body, face-to-face and real-world unpredictability simply doesn't exist. The lineup at the border, the flat tire or the runny nose don't compute on this screen of reality. And that's the point, that's why it can set such a quickening pace, shortening production cycles, compressing time and space: because any squeeze on the people and the natural environment is out of sight and out of mind. It disappears.

Behind the computer screen, too, everything connects with everything else, so fast that it builds its own momentum. The two work in tandem: the dematerialized, dis-incarnated symbols and the speed of light at which they move. Yet more and more the two are now defining and driving reality off-line, turning places into interchangeable portals and people into units of labour, consumption, credit or accreditation. The fast drives out the slow

43

everywhere, from the market for new products to the latest in television and Internet news. Decisions are driven by how fast you (or a colleague or competitor) can finish the (computer-aided) design, how fast you can order the supplies and assemble the parts (using on-line inventory and ordering systems), how fast you can have them out the door, how fast you can make final delivery.

Two centuries of standardization and compression, coupled with progressive acceleration in mobility, have shrunk the world into a miniature model of itself, more real as a hand-held image than it is as flesh and blood. Moreover, people are now pushing the buttons of this miniature-model world as if you *can* hitch this unit to that one, this module of investment to that contracted-out business service, in shorter and shorter windows of time and faster and faster flows, without consequences outside the immaculate model where symbols flash and fuse under the corporate flag or logo. With the flip of thinking that is associated with terms like "the new economy" and "globalization," particularly as this has been promoted as "globalism," an ideology extolling fast, efficient bottom-line thinking as the best way to run the world, the focus now is the symbols representing these interchangeable container-units, including bar codes on the sides of freight cars and container trucks. These can be tracked and controlled from afar as pure information, pure symbols, pure data. Wisely or not, this is the only reality these deal makers and managers see. Any realities associated with an overheating planet, or an overextended body or institution, are left in the dust. The squeeze of time and circumstances, and any coffee-shop bitching about the stress involved, become residual. They're hardly real at all.

IT'S MIDNIGHT, and we're in Montreal. Wayne hauls on the wheel, pulling his rig off the highway. He navigates the nearly empty streets, the stop lights lonely in their superfluous sequences, and drives into the depot. We could be back in Toronto, the freight terminal is so identical to the one we left behind, with its cookie-cutter loading bays down each side of each warehouse. The particulars of where we are hardly register. Wayne backs the container into its appointed slot, grabs his clipboard and heads inside. I follow him, listen, smiling as he greets familiar faces. "Hi. How are ya?" "Not bad, not bad." As he goes about his business it strikes me how insignificant his conversation is. It's not just having little time for more than chit-chat. There's little scope for going beyond this because the business at hand is already in hand. It's all laid down in the computer dispatch for his outgoing load, a symbolic replica of the container waiting outside for the click of his hitch, an adjustment of the pressure going to his air brakes, the power flowing to his backup lights. The crux of it has already been negotiated and decided by a bunch of symbolic analysts and intermediaries, in a conversation among their data sets and machines. Wayne could probably print off the work order inside his cab if he had a printer tucked in there. Then he'd have no need to even come inside this nearly empty depot, except maybe to use the bathroom. Night staff could be replaced by a call centre and the washroom by another unit of GE (General Electric) Modular Space.

TRUCKERS LIKE Wayne Bates are almost surrogates for the rest of us who, if not working through the night, still are just trying to make that window on delivery that looked so easy when we signed on. Truckers are surrogates for our isolation, families as close as a cellphone call but otherwise far away. These drivers dramatize the overextension involved and possibly, too, the imminent crash. There are tens of thousands of them out there, now.

Between 1994 when NAFTA came into effect and 1999, U.S. trade with Canada nearly doubled to $410 billion a year. During that period, trucks hauled more than 61 million metric tonnes of cargo. They also put out 38,000 kilograms of nitrogen oxide, 20,000 kilograms of carbon monoxide and a staggering 5 million kilograms of carbon dioxide a day. Nitrogen oxide is a precursor to ground-level ozone, or smog. Carbon dioxide is the primary component of the greenhouse gases that contribute to global warming.

The lethal emissions add to the story of what happens when material reality is compressed out of the picture. Here it's the devastation of Earth as it has been transformed from a living and life-giving place into mere transportation corridors dotted with portals of production, distribution, sales and consumption. The buildup of these poisons is being allowed to continue partly because we're bereft of both a citizenry organized and committed enough to articulate the danger this pollution represents into an urgent international agenda and a language with which to express this as real. The voice of the body sensing a palpable danger is seldom heard. The voice of direct experience has been progressively silenced. And it's likely to remain that way as, individually and collectively, we carry on because the cellphone's ringing or the computer's beeping or there's a delivery window to make.

IT'S PITCH DARK in the cab except for the glow of Wayne's cigarette, the lights on the dash and the single red light on my tape recorder. We hum along in silence for a while, more golden arches flashing by, a sign for another Flying J truck stop. They're all markers in a skein of compressed space and time, and every night Wayne works it through his hands afresh, as a string of numbers. He asks me to guess what's in his load out back. "Illegal immigrants," I think, but say nothing. It's Styrofoam, he announces happily. The entire container is packed to the roof with little trays bound for some fast-food outlet somewhere. It's like carrying a load of feathers, Wayne grins. And with no headwind, the numbers should come out pretty.

"It costs you 54 cents for every mile you drive, that's for fuel. Then you add on 11 cents a mile, that's for maintenance, okay? That's 65.9 cents; that's what it costs me to drive my truck down the road. Now, I haven't paid for my truck. I still have to pay for my truck! And that's no wages. That's no benefits. That's no workman's compensation!

"Then you wonder why some guys cheat in their logbooks to get extra work! Well, some guys, they have to. Sixty hours a week isn't enough, okay? To make those payments. You're squeezed."

TRUCKERS are expected to keep an accurate account of their running time in the logbooks they carry with every load. They're required to record the time they leave the depot, the time they stop for coffee, the time they start up again and so on. The time they record, however, isn't necessarily accurate. The logbook might dutifully honour all the limits to truckers' hours that have been legislated into effect for the safety of truckers and other motorists, though deregulation in the trucking industry has tended to roll back those limits. Still, close inspection of the documented numbers, in the context of the distances covered in that time, can uncover challenging discrepancies. (A 1996 U.S. Department of Transportation probe into trucking safety found that over a four-month period ending in January 1997 truckers for McCain Foods falsified logbooks an average of forty times a month, driving far more hours on the road than the laws in either Canada or the U.S. consider to be safe.)

"Sleep deprivation, it's the issue no one talks about," one trucker told me. "You cut your sleep time short all the time. It's the only way you can do it." That's the way the industry is run these days. It's an economy of flows, I think, and a crash culture: a culture, an ecology, in the throes of crashing.

IT'S 2 A.M. We've just loaded up with more fuel and another big mug of coffee, and Wayne's opening up a bit more. He's telling me about the tricks of the trade, about sneaking around the "chicken coop." In other words, the weigh scales and inspection stations that dot the sides of the highways.

The one at Whitby, he tells me, is one of the easiest to get around. "If it's open and you're doing something you shouldn't be doing, you get off the highway right in front of it, and you come back out on the other side.

"Now, they patrol that real good there. So, well, the guys that want to get around there, they just get off at that ramp, and they go down and they buy themselves a coffee, because there's a coffee shop right there. And when they go back on, they've missed the scale. But don't get caught without goin' in and buying that coffee, or you're euchred. Because if they see you get off and just get back on, they'll know what you're doing. And they can bring you right back, and they can charge you for trying to avoid it.

"And some guys go around because they can't afford to waste the time being inspected. Or they can't take the chance (because there's an obvious lapse in their logbook or they've got something else to hide)."

I wonder whether Wayne himself has done much of this cheating, just as I wondered earlier, when he was talking of road rage affecting others, whether he has ever snapped. But I don't say anything; it would seem rude, a betrayal of the trust between us, somehow, to put him on the spot. Instead I sink into the cozy, companionable silence, the staticky squawk of the CB radio, the drone of the all-nighter trucks.

It's 3 a.m. I crawl into the bunk behind the seats and melt into the horizontal position, though I revive briefly to take in Wayne's miniaturized world of creature comforts. It's a complete life-support system shrink-wrapped to fit this small, warm space. I think of an astronaut on a space walk, the life-support (everything!) packed around him in his space suit. Maybe soon Wayne will have oxygen in here, I think, or at least an air-filtering system for the pollution on the road.

When I wake up, Wayne has the windshield wipers going and some of his gospel music on the tape deck. He's getting clear of a patch of snow squalls, he says, and Toronto's on the horizon.

I buckle myself back into my seat and sit there dozily watching an incipient dawn bring landmarks back into shape. He just might make it, Wayne tells me. His goal is to get to the depot and back to his rig before the sun crosses the horizon. If he can shut his eyes before the sun comes up, bingo, he's straight off to sleep. Otherwise, his biological clock goes off and he's stuck, miserably exhausted and only half-asleep. He should make it okay today. He's well ahead of morning rush-hour traffic, though at 5:45 the commuters have started to stir, getting a head start on their own just-in-time day.

Two

STRESSED OUT AND DREAMLESS

*"It takes all the running you can do to keep
in the same place. If you want to get somewhere
else, you must run at least twice as fast as that."*
LEWIS CARROLL, Through the Looking Glass

*"We are moving from a world in which the big eat
the small to a world in which the fast eat the slow."*
KLAUS SCHWAB, EXECUTIVE CHAIRMAN,
World Economic Forum

IT'S NOT THAT I hit the floor running, but there's a certain momentum to my getting up in the morning. I savour the flick of the switch as I turn on each appliance: the kettle heating while I take a shower, the fax sending a message while my computer warms up. Tea mug in one hand, mouse in the other, I click to make the connection. While the modem dials into the server, I reach for an unopened piece of mail. I think of making a phone call, but the line is tied up. (Note to self: get that second line next time a telemarketer offers a sweet package deal.) I take a sip of tea and hear the burst of static that signals I'm on-line. Tea mug down, I move the mouse, positioning the cursor so that it's ready to click my e-mail onto the screen. With my other hand, I pick up an empty envelope and chuck it in the recycling box.

As the e-mail trickles in I sip my tea, scan the titles and answer the easy ones. I click and forward this one, file that attachment for later. I hear the bells and whistles announcing an incoming fax in the other room. I reach for the pile of unread papers on my desk and pull my daybook out from under one of them. Then I position the cursor over the e-mail "delete" box and mow my way down the line. Delete, delete, delete. I feel a sense of purpose, as if I'm accomplishing things. Delete, delete, delete, delete. What was that? I catch myself deleting the notice of a gallery opening for an artist friend of mine and think that maybe it's important to him that I be there. Important for me, too. I consider undeleting the message, then think, why bother? I have no time.

I sift through my mail, write a cheque for a bill I haven't yet transferred to a pre-authorized payment plan, and by the time I lean back and grab my mug the tea's gone—all gone—and I can't remember having drunk it. It's as though it never happened, or I wasn't here enough to experience it. I'm connected all right, yet strangely disconnected at the same time. I look down at my multi-tasking hands. It's not that the right hand doesn't know what the left hand is doing; it's that they're competing! I'm a one-person time-motion study, a perpetual-motion machine.

Oh, I can make myself stop, slow down and be fully present to what I'm doing. I can take a break or make another cup of tea, or coffee, and savour it. But before I know it, some part of my mind has slipped off to think of this or that project, to replay what I've said or done, to rehearse what I will say to this or that person, or even to plan what messages I'll leave as voice mail.

On good days, I'm Proteus, the shape-shifting god of Greek mythology. On others, I'm more like Alice on the Looking Glass's moving chessboard, running like crazy just to keep up. The to-do lists are always scrolling away in my mind, seldom leaving me time to just be quiet and tuned inward, in touch with myself. I'm probably what cultural theorist Mark Poster has in mind when he writes of the dispersed subject of the postmodern new world order; I'm scattered, yet I carry on like so many others I know, wearing my fatigue and memory lapses like a badge of honour.

THE DISAPPEARANCE OF LEISURE

What happened to the leisure society and to the social-welfare state that tacitly supported it as it balanced economic efficiency with sustaining family, culture and community life, that talked of social indicators and not just economic ones? Something fundamental has changed. In 1991, the majority of Canadians worked less than forty hours a week. But by 2001 36 per cent were working more than forty-five hours per week, and another 25 per cent were working over fifty hours. At the same time, 38 per cent described themselves as "highly stressed."

Much of the explanation stems from the economic crisis of the early 1970s. The solution that emerged most strongly and swiftly from that period was the neo-conservative business model: a globalized reconfiguration of capitalism in which container-like modules of financing, management, production, marketing and distribution are combined and recombined, using the global digital networks of an emerging "knowledge economy" to outsource all the work and much of the investment in the physical accoutrements.

Originally dubbed Thatcherism or Reaganomics (for their initial champions British prime minister Margaret Thatcher and U.S. president Ronald Reagan) before settling into more generic titles like "globalization" or "the post-Communist new world order," these policies ranged from cutbacks in government and public spending to deregulation in business and enhanced "free trade" and competition. Besides promoting business methods for maximizing efficiency and global competition, they attacked public services as inefficient and the social-welfare state as a costly drain on the economy. As a result public policy was driven into retreat, as were labour and environmental standards. A concern for balancing society's, nature's plus business's priorities was replaced by

fast-tracking performance and economic indicators, under the enabling eye of the World Trade Organization.

The medium through which much of this ideological shift took
place was also key. The whole infrastructure for doing things in society was transformed from what has been called the "Fordist"
model of industrial assembly lines and fairly rigid organizational
bureaucracies to a "flexible" and "agile" post-Fordist model organized around computer-communication networks.

The Internet is the hub of this new economy. It's even referred
to as a "lifeline," with companies using it to reposition production
and to calibrate personnel, inventory, customer service and pur-
chasing to a "just-in-time" last-minute precision. Lines of communication used to be an adjunct to actions on the ground, much as
telegraph lines running alongside railway tracks were an adjunct
to the movement of freight. Now, however, these once separate
lines have become smart and almost seamlessly interconnecting
digital networks. These global networks have become where the
action itself is happening, which has made them the hinge in the
shift of paradigms to a post-industrial information and knowledge
economy centred on symbols. The networks are both a medium
for transmitting information on what's to be done and an operating environment in their own right.

The next step was to legitimize the transnational market
expansion this permitted, through successive rounds of trade-
liberalization talks that led to sectoral "free-trade" zones through
agreements like the North American Free Trade Agreement.
Canadian production and distribution systems, for example, were
reorganized to integrate with those in the United States and
Mexico. These new systems were not only faster and leaner but
more flexible, with new design specifications and operating
instructions relayed to them over the Net and even directly   53
into production machinery that could then be reprogrammed to
take new job opportunities. The key was not so much robots and
other automated production equipment, but rather connecting

management-information systems to any and every part of an enterprise and contracting out work to any other part of the world, thereby compressing space between points to virtual insignificance. Now that globalized management is being extended to services as well as manufacturing and distribution, trade-liberalization talks such as the General Agreement on Trade in Services are laying the groundwork for global free trade in education, training and even health care.

Some people were laid off as their work was contracted out to companies specialized in expert knowledge and remotely managed through integrated management- and financial-information networks. Some jobs were lost as on-line data-processing and "expert" software took over from humans who had previously processed lab-test results, expedited bank transactions and updated personnel files. Still other jobs were lost as industrial society was reconfigured as modules of goods or service production that could be outsourced to Barbados, China, India or to maquiladoras, factories along the U.S.-Mexico border taking advantage of cheap labour and lax environmental and labour standards but exporting their products to foreign countries.

This has resulted in a scramble for fewer—and less skilled— jobs reflecting what economist David Ricardo called the "iron law of wages," which decrees that wages will tend to stabilize at around a subsistence level. (Today the law is being applied globally, not locally, with the result that, under competition from low-wage China, inflation-adjusted Mexican minimum wages have dropped 21 per cent since 1994 to what is for many a level *below* subsistence). Now, whole "factories for hire," special-purpose or multiple-purpose production units equipped with state-of-the-art technology and production teams, bid for contracts and pull them in from around the world if the systems and skill sets they offer fit the design specifications, the budget and the production schedule.

Companies routinely outsource administrative-support, human-resource management and accounts-payment work, plus a

54

whole range of customer service. It's no longer New Brunswick with its few hundred call centres (or, as they're called now, "customer-contact" centres) competing against Ontario with its few thousand. Now the real competition comes from India, where customer-service reps take on "aliases"—Americanized names like "Polly" (instead of Pratibha) and "Derek" (for Singh)—and are trained in "accent neutralization" so they can pass as virtual locals to the North Americans (and sometimes Britons) they call during their often ten-hour overnight shifts (mapped to local daytime hours). The business press claims that business-service outsourcing, or "offshoring," will soon match the trend in manufacturing. The competition for jobs, especially those paying a decent income, is likely to continue.

The leisure society is also eroding because people are working longer hours for less money. Between public-sector cutbacks and downsizing plus layoffs and jobless growth, the employment profile that has emerged in Canada and throughout much of the Western world features sharp and deepening inequalities, the so-called "good" jobs/"bad" jobs divide. At the top of the pay scale (the good jobs) there is a super elite of the working rich, followed by a greatly diminished and hollowed-out corps of professionals, paraprofessionals, administrators and managers in the middle who, often now, are self-employed. At the bottom of the pay scale (the bad jobs) are a large number of often permanently part-time employees working automated machinery in call centres, warehouses and factories and in business and retail outlets of all kinds. As an example, of the more than half a million jobs created in 2002, 40 per cent were part-time and another 17 per cent were for self-employed contract workers. An estimated 8.1 million Canadians are working for less than $25,000 a year; of these, 5.5 million are being paid less than $15,000, which is hardly a living wage.

Given the relative scarcity of "good" jobs, people who had taken full-time employment for granted suddenly felt grateful and were willing to express their thanks by voluntarily staying late or taking work home with them at night—at an average rate of four hours a week, according to a Health Canada study. People in the new "precarious" forms of part-time and short-term contract work struggled not only to maintain what temporary positions they had but to work toward something more secure. They took on extra work to build their reputation for loyalty and on-line courses to build their credentials. Today working on-line, using e-mail, voice mail and various point-and-click shortcuts has become a popular survival tactic for doing more with less. It also has become a conduit through which new demands and expectations are being visited onto a range of people, not just in the private sector but throughout the public sector as well. In today's wired universities, for example, academics cite increased demands and expectations, from administration, students, and even colleagues around the world, as the biggest reason for rising workloads and stress.

The media have played up the buzz of being on-line, the ultimate freedom it allows, the awesome megabyte power and extreme individualism in the wild new frontier of the Net. As well, after a decade or two of deficit-cutting and restructuring, e-commerce, cellphones and on-line culture were a way to lighten up. At the time, a business-friendly press fawned on celebrity CEOs, depicting them as front-page superstars, always laughing, always wearing open-throated shirts and living in penthouses designed for multitasking. The ads accompanying these articles sold high-end workstations as lifestyle and wireless hand-held computers as gold-key entrees into the zone of business-opportunity flows. By 1999, more than 50 million people in the U.S. and Europe were considered "mobile professionals" supported by cellphones, laptops and personal digital assistants (PDAs). By then, too, most Internet traffic was for business, and the majority of home computers and on-line connections were being used for work and for upgrading education and skills. Dreams of a leisure society were eclipsed by a jazzy upgrade of the work ethic coupled with career-enhancing conspicuous consumption. Success, even survival, now depends on every-one from self-employed entrepreneurs to corporations and their employees being able to live "in a permanent state of emergency, bordering on the edge of chaos."

WHO'S CAUGHT IN THE SQUEEZE?

Not everyone is under pressure. The people at the top of the hierar-chy, the so-called winners of the new economy, are sometimes also called the new "time masters." They're the "one-dayers," the people who can see practically anyone they want to within a day, leaving others scrambling to respond by reprioritizing and rescheduling their own day. One-dayers have the most discretionary time; in fact, it defines their status and their power. To stay where they are, they also use their time with consummate skill, surrounding them-selves with the best in on-line technology and the brightest people, whom they keep at their beck and call around the clock.

The people at the bottom of the hierarchy are under their *own* pressures. Many are squeezed by low wages, split shifts and often boring, sometimes brutal work conditions. Many of these work in the burgeoning realm of call or customer-contact centres.

"They hire us to be yelled at," one customer-service representa-tive told me bluntly. She understands her job to be acting as a pres-sure-release valve for customers finding bugs in systems that they are inadvertently helping the company to fix for the next genera-tion of product. The verbal abuse contributes to the high turnover rate in such jobs. It also contributes to a certain studied indiffer-ence, or dissociation, a refusal to engage beyond the level of the limited scope of the prescribed task. In particularly sweatshop-like settings with tortuous split-shift assignments, some employees have tried to form unions—and some have succeeded—but often the strategy is simply to "move on," much as the employers do.

The people in the middle are under the most pressure—or rather, the pressure they're under concerns me the most here. I'm

57

referring to the professionals and the managers who are generally counted on to run things in society and keep our social institutions healthy. I'm particularly interested in their experience because, traditionally, their jobs have required that they engage as whole people in their work—treating patients, educating students, facilitating work groups. They've often worked in collegial groups and autonomous institutions and have taken such things as continuity and discretionary flexibility for granted. The stress these people are under is a result not only of the volume and the pace of their work but also of a fundamental shift in its nature. It might still be nominally centred on human relationships but increasingly these are mediated by on-line technology. The stress here, then, includes people being drawn away from the realities in front of them and into a new, almost virtual level of reality that is remote and based on abstract symbols that may or may not make sense to them and yet to which they are increasingly held accountable. Their stress takes this book well beyond issues of personal health to consider how as a society we might be spiralling so far down into dysfunction and dis-ease that, as Jane Jacobs, author of *The Death and Life of Great American Cities*, suggests, we could be entering another Dark Age.

What happens when the alienation and its compensatory passivity, dissociation, apathy and indifference that's so taken for granted in factory jobs move farther up the food chain of our society?

UNDERSTANDING STRESS

By the late nineties, the statistics show, the pressure was taking a toll:

In 1998, the Canada Health Monitor reported that Canadians were three times more likely to complain of stress in the workplace than of any other health problem. They cited pace of work as the most common cause of complaint.

That same year, the International Labour Office in Geneva labelled job stress as a chronic disease of our times. Other reports called it an epidemic.

Two years later, a millennium study by the Heart and Stroke Foundation reported that more than 40 per cent of Canadians over the age of thirty described themselves as "often" or "almost always" overwhelmed by stress.

In Japan, the term *Karoshi* has officially drawn a link between more than 10,000 deaths a year (mostly heart attacks and strokes) and the over-demanding work environments that are a hallmark of the country's continuous-improvement management regime.

A survey of American and British workers reported that workload and deadlines are the Number 1 cause of stress in the workplace.

A 2001 study by the Families and Work Institute found that nearly a third of U.S. workers now feel "overworked" or "overwhelmed by what they have to do."

In Canada, a landmark 2001 Health Canada–sponsored study of nearly 32,000 people in the public and private sectors found widespread symptoms of stress, not only from long hours of up to sixty hours a week, including regular after-hours work at home. They're also squeezed by the "struggle to juggle" work and family-life responsibilities. The rising standard of living, plus the lack of adequate parental-leave policies, except in Quebec, has more and more parents of very young children working as full-time as possible. Forty-seven per cent of the people surveyed reported that they participated in "family-time" events like sharing a meal, doing things with their kids around the house or going out as a family only once a week. Twenty-seven per cent said they "rarely" engaged in these activities.

People are working longer hours and also working more intensely, juggling many tasks and files at once, responding to new demands, including for accountability. In a survey of social-service workers in Ontario, 65 per cent reported there isn't enough time to complete the necessary documentation. Sixty-three per cent reported working two and a half hours of unpaid overtime a week to complete it, and 12 per cent are working an extra five hours and more a week.

So workplace stress arises from working longer hours and harder; it also comes from trying to stay ahead. Work-related learning has morphed from a blip on the timeline of life to an ongoing taskmaster. It now absorbs five hours a week for most people surveyed in the 2001 Health Canada study as more and more people change jobs and careers and, for the growing ranks of the self-employed, get used to new project and contract teams. Just commuting to and from work—including to jobs many have had to look farther afield to get—is taking up more time and creating more stress.

Finally, another factor surfaces in a lot of the smaller studies: loss of control as decision-making is centralized and local discretion and latitude are curtailed. A 2002 study of senior executives in the federal civil service noted that although a 1997 report warned that reduced latitude in decision-making was contributing to stress severe enough to trigger heart disease, nothing had been done and the health of these people had deteriorated. In the U.S., researchers have found that a lack of supportive supervisors and autonomy on the job are bigger causes of stress than long hours.

To better understand what all these numbers really mean I contact Linda Duxbury, one of the co-authors of the Health Canada study and a colleague at Carleton University, where I teach part-time. We meet at her home office, where we sit back over a lovely, rich cup of coffee to discuss the causes of stress. She begins with her own checklist: (1) running short-staffed has become the new organizational norm, (2) technology keeps changing and advancing, not just in one area (such as computers) but in many others at the same time, and (3) the Internet and, with it, the ability to work and to be in touch with others asynchronously is leading people to be "on" 24-7.

"That has really changed everything," she says. She looks back to her computer. "It's the communication, the technology in combination with the ability to work any time, anywhere, and contact anybody, any time, anywhere!"

I take up what seems to be the corollary to this thought: how much this puts people out of sync with each other, even with ourselves, because we spend so much time facing various screens rather than one another in shared time and real space.

"Yes," she says, "it's a whole new model, and we're not skilled at it yet. We haven't developed an etiquette around it."

"Etiquette?" I say, trying to hide my dismay at such a narrowly individualized response to what I see as a social environment being rendered toxic by an inhuman pace and dangerously irresponsible by an abstracted, disembodied form of presence.

Linda laughs and sets down her thermal mug. "I'm not a public-policy sort. I'm an organizational-policy sort, and an individual-coping sort." She's a professor in the Sprott School of Business, after all.

I consider all the signs of deteriorating civility in society, everything from road and checkout rage to ubiquitous cellphones, plus the research probing its causes. Pier-Massimo Forni, co-founder of the Johns Hopkins Civility Project at Johns Hopkins University in Baltimore, highlights three factors: anonymity, stress and narcissism—what he calls a "narcissistic cage." I ponder these. Anonymity is "isolation at work caused by technology" and the "impersonal" nature of e-mail and other new forms of instant, technology-mediated communication. Stress includes corporate downsizing and overwork. Narcissism speaks of a popular culture built on being self-centred, and of good manners not being part of a child's upbringing any more.

Linda tells me that some Ph.D. students she's supervising are researching virtual digitally networked work teams to help identify what's needed for them to connect with each other personally. She adds, "Because we know that if they don't get together face to face fairly quickly, they in fact will walk all over each other because they don't see each other as people."

Instead, what they see are all the insensitive, decontextualized, disembodied demands coming at them through all the portals of their existence. Even if these messages are coming from people they know and might even like and respect, I muse, the demands seem insensitive, almost anonymous, because they're just these abbreviated bits, or standardized forms and data sets.

"That's it," Linda says. "That's the Number 1 problem: rising demands and expectations." She grabs her computer mouse to find more data to support this finding. "Keep going, I'm listening," she says, because I stopped talking when she broke eye contact. I watch as she clicks a few keys to send the relevant chart to me by e-mail.

One of her graduate students has dubbed all the new gadgets— the laptops, cellphones, BlackBerries, notebooks and other handhelds—"work extenders," she tells me.

I ask if she's read Marshall McLuhan's description of technology as "extensions of man" and simultaneously "auto-amputations."

"But we're not conceptualizing it like that," she says. "A lot of people see the technology now as a ball and chain. The technology is an extension of work, not of your ability to have a life."

I ask if she'd go as far as McLuhan did in his warning that the scale and pace of technology could actually amputate the rest of our lives by numbing our sensibilities.

"Well, it's certainly getting toward it. You have a kind of hierarchy of things you have to do. And in many people's hierarchy, leisure and time for themselves is right at the bottom. So you have to amputate. If you're filling your day with work and secondarily filling it with looking after your children or your elderly parents, then what falls off the board is time for yourself."

Discretionary time and space. They're taken-for-granted aspects of middle-class entitlement. They're a defining feature of liberalism, and even a central tenet of privacy, if you understand privacy as the ability to define the boundaries of our own lives in more ways than the obvious ones of keeping the state out of our bedrooms and strangers off our phone lines. But beneath the mystique of being wired to go anywhere, any time, a subtle change has occurred, and we can no longer take having discretionary time and space for granted.

Academics like Linda are among the most overworked in the downsized, globally hot-wired new economy. A British study reported in 1994 that university professors were working fifty-nine hours a week, and nearly sixty-five hours if they were women. In Canada, the study I helped initiate among Canadian academics had concluded only its pilot-study phase when I wrote this book. Still, the results were consistent, and alarming. Not only are academics working long hours, including evenings and weekends, they're reporting high levels of stress that include headaches, insomnia... and also strained relations with colleagues and even friends and family.

"That's in the next [Health Canada] report," Linda tells me, speaking of the massive toll stress is taking on people's health, physically, mentally and psychologically. She reaches for her mouse again. Do I have high-speed Internet access at home, she asks. She couldn't live without it, not just for corresponding with her graduate students—and she takes on about eight Ph.D.s a year plus three masters students—but to relay these big files around to co-researchers at other universities and at the Canadian Policy Research Network with which she is affiliated.

Later, when I download the advance copy she's graciously sent me, the evidence from the 32,000 Canadians surveyed matches the symptom profile other researchers have been assembling over the

63

past five to ten years. It's not just the obvious signs of stress: the chronic headaches and chronic exhaustion, stomach disorders and muscular-skeletal breakdowns, the hypertension and heightened risk of heart attacks. It's also the broken night's sleep, the insomnia, the short-term memory loss, immune-system disorders including chronic fatigue, the suppressed ovulation in women, plus the anxiety and depression characteristically associated with a sense of helplessness and hopelessness about one's situation in life. The cost in health-related absenteeism and disability claims is $5 billion and rising. So is the cost of various medications, including sleeping pills and tranquilizers: an estimated 17 million Americans are taking the antidepressant drug Prozac, making it the second most-prescribed medication in the world. (Given what antidepressants do to the libido, no wonder Viagra has become so popular and, wink-wink, is described as the perfect weekend pick-me-up.)

Delving further into the connection between stress, depression and prescription drugs, I read that the World Health Organization estimates that mental-health disorders, which include depression, are well on their way to becoming the second-largest cause of disability and death. In 1995 more than 1.5 million Canadians sought treatment for depression, making it the single most common psychological condition seen by family physicians. By 2001 nearly 33 per cent of working Canadians were suffering from depression, compared with 14 per cent in 1991.

I look up from the screen after taking all this in, suddenly realizing that what's missing are the living (or trying to live) people behind all those statistics. What's gone missing is what this means, what this actually feels like as human experience. To think that I spent two hours in the company of another person who is possibly experiencing at least some of these symptoms as I am, and yet we never touched on the personal even as we sat there face to face! There were so many missed opportunities, I think now. Just talking about our matching mugs of strong coffee, we could have gone personal from there. How many of us, I wonder, keep going literally on the strength of the designer coffees because they pack a caffeine jolt that Health Canada recommends as the daily maximum? I could have told Linda about my daily dose of ginkgo biloba, a natural supplement to boost my brain; I could have 'fessed up, too, to the nightly dose of valerian-root extract to slow me down enough to sleep.

Multi-tasking. Why didn't I stop Linda when she'd break off conversation to click and find something in her database? We could have taken a hard look at what it means to be focussed briefly and intensely here, then there, and to participate simultaneously in a conversation. I might have asked: "Have you ever found yourself locked in that processing mode, getting almost twitchy if there aren't demands coming at you, multiple things to be done all the time?" What's lost when we gain all that stuff of efficiency, I wonder.

"Keep going," she'd said when she reached for her mouse and I broke off talking. I might have told her that it left me feeling a little sad and lonely, as though she'd deserted me, if only briefly. We could have explored what it means to be present yet constantly projecting forward, again in task-identification mode. For me, it's as if I'm never fully anywhere, never dwelling in the amplitude of the moment. My senses are blinkered, effectively blind to the stuff in my peripheral vision, deaf to such subtleties as a shift in the rhythm of the conversation—closed off from the surprises and the fecund play of lateral thinking that is key to original thinking.

We could have talked about discretionary time and what it means personally to lose it. She once laughingly described her family as a household of overachievers. Matching her own considerable career, her husband is an associate dean. In addition to school, her daughter is into competitive water polo and competitive soccer, plus yoga. But they schedule downtime, Linda told me. Together she and her husband do two karate classes and one yoga class a week. In the summer they take a whole month off as a family, slowing down to the point that they play board games and just hang out. Last year they started their holiday with a trip to the Galapagos Islands, she told me. This year, Fiji. That's wonderful, I think, but besides being an individualized solution that not everyone can afford, it's not the same as having a free-play sense of time and autonomy in the daily social environment.

Linda and I had the discretionary power and time in our visit together to have discussed our inner doubts at the trade-offs people are making as they adapt, plug in and log on. Yet we stayed in the zone of smart, fast talk that is so current these days; we stayed in the cell of our professional selves. But why? Why did I remain silent? Partly, I think, because of our age difference. It's less than five years; still, I think, at fifty-five maybe I'm too old for this game. Partly, too, because we look at the world differently. As she told me, she's an "organizational-policy" and "individual-coping" sort, whereas my "sort" looks at the larger human and societal picture and the public policies required to advance the collective public good. The language of data sets, economics and administration that currently dominates policy discussions subtly clips and mutes much of what I have to say. But there's more to my silence than that. It's the stress itself that leaves me chatting away on the surface of things. The stress robs me of the confidence and inner presence to feel what I'm experiencing strongly enough to speak of it. That's the epidemic I want to talk about!

TRASHING THE BODY

The concept of stress, as we know it, dates to the mid-1800s, when the English medical journal *The Lancet* described a form of nervous fatigue resulting from the "incessant shifting of the adaptive apparatus" of perception as objects flashed by the train window. After two world wars and confrontations in Korea and Vietnam, over the course of which it was called "shell shock," then "combat exhaustion," "battle stress" and "combat fatigue," the shock or fatigue involved was understood to be a generalized psycho-neurological ailment, a state of dis-stress or dis-ease, of being unhinged from

66

one's normal basis of equilibrium. By then the stress associated with combat had been studied to the point that preventive action had almost eliminated it. Two factors seem to be critical. First, battle stress tends to set in after prolonged, sustained exposure to continuous bombardment and continuous crisis. Second, it also results from a constant experience of the unexpected, of things being beyond one's control and, with this, a growing sense of being overwhelmed and unable to cope. The remedy seemed to be to rotate people away from the front lines before their coping capacity broke down, though this didn't become army policy until after World War II. By that time, too, a third factor had been understood, namely the danger of soldiers being alone and cut off from each other. One key to preventing stress was "the sustaining relationships with others of the battle group."

In the past several decades there has been much research into the physical effects of stress on the body. Hans Selye, one of Canada's leading experts on stress, has stated that there is always stress in life. The question, therefore, is how well does the body respond. Selye's work asserts a normal state of inner balance and equilibrium to which the body seeks to return. When stress first hits, the body's defences initially go down. However, if the stress continues and the body can adapt, resistance and defence levels not only recover but improve. Hence the "high" of being "on," or "in the zone." The trouble is, there are real limits to our ability to keep coping and adapting. There are limits to how long the adrenalin pump can stay on without the risk of it running dry if it can't complete its normal cycle, which includes a necessary period of calm. Although stress adaptability will vary from one person to the next (depending upon, among other things, personal background and age), eventually what Selye calls "adaptation energy" becomes exhausted. Healthy stress, or "eustress," becomes distress. Adrenalin fatigue sets in, numbing people's sensibilities while, behind the frayed and frazzled nerve endings, vital organs, including the immune system, start to sicken. "In short," Dr. Archibald Hart wrote in *The Hidden Link between Adrenalin and Stress*, "People in our age are showing signs of physiological disintegration because we are living at a pace that is too fast for our bodies. This is the essence of the stress problem." According to McGill University researcher Sonia Lupien, chronic and unpredictable change, plus a perceived loss of control over what's going on, are also important causes.

Social scientists and occupational-health researchers have documented links between stressful work environments, including long hours, and increased levels of stomach medications and tranquilizers. One study, pursuing the link between prolonged overwork and heart attacks and strokes, found that long work hours increased "the burden on the cardiovascular system of workers with normal blood pressure as well." Not only did these workers' blood pressure go up, their heart rate also tended to increase. Another study has documented a link between stress and increased levels of serum cholesterol in the blood, a danger sign for heart attacks. Equally, the studies emphasize that it's the chronic nature of being "on" and subjected to new things grabbing at one's attention that do the damage.

To recover to normal cycles after prolonged stress, circadian rhythms need up to five days. Otherwise, stress-hormone levels stay up, even on so-called rest days. The immune system becomes exhausted and endorphins depleted, lowering a person's tolerance for pain and discomfort and increasing their anxiety to the point even of triggering panic attacks. In the words of one researcher, "The mind becomes obsessed with petty problems, and thoughts keep going 'round and 'round like a stuck record."

Scientists across a range of disciplines have found that stress not only throws the body's pH levels out of balance but causes a rise in cortisol and a drop in salivary immunoglobulin A. Cortisol, which is pumped out by the adrenal gland, is considered the "negative" coping hormone because it causes digestive problems and sleep disturbances. It's also been particularly linked to work environments where "active control is difficult" or "demands cannot be adequately met." (Conversely, when people have the leeway to work at a pace that's in sync with their own physiological rhythms, that is their breathing and heart rate, not only are they less fatigued, they are more efficient.)

Thirty years' worth of animal research on corticosterone (the animal version of cortisol) shows that it can also atrophy the brain's hippocampus, which is an important site of learning and memory, leading to short-term memory loss. Cortisol suppresses the immune system, as well. Immunoglobulins affect us as dramatically. They are a group of structurally related proteins that function as antibodies, and immunoglobulin A is the most predominant one found in fluids such as saliva, breast milk, nasal, gastrointestinal and bronchial secretions. As such, it's considered a key immunological barrier. As it goes down, so do the body's defences against disease. A range of diseases, from adult-onset asthma to rheumatoid arthritis and even Alzheimer's, have recently been linked to stress and to its damaging effects on the body's immune system.

So why don't we just take an aspirin or a warm bath and go to bed? Our overtaxed bodies, minds and immune systems can for the most part repair themselves through sleep. But many people are so stressed and overworked that even when they lie down they can't get to sleep.

SLEEP DEPRIVATION AND DREAMLESSNESS

Downtime and sleep aren't just personal things. They're not indulgences of the weak and the less-than-fittest of the species, as Margaret Thatcher suggested with her celebrated four hours' sleep at night, and Microsoft chairman Bill Gates and his cohorts in Silicon Valley assume with the boast that "our programmers don't need much sleep." Sleep loss dumbs us down to the point of dangerous disability that has been likened to driving when drunk. Research has linked "sleep stupidity" to some of the worst disasters in recent history—including the catastrophe at the Chernobyl nuclear power station, the meltdown at the Three Mile Island nuclear power plant, the oil spill caused by the foundering of the *Exxon Valdez* and the explosion of the *Challenger* space shuttle. According to University of British Columbia neuropsychologist Stanley Coren, the real cause of the *Challenger* tragedy was sleep stupidity, as National Aeronautics and Space Administration (NASA) personnel scrambled through fourteen-hour days for up to twenty-six days straight to meet a launch target and didn't pay attention to warnings about the faulty O-ring seals. "We are a dangerously sleep-deprived society," he says.

The statistics bear out his claim. The National Sleep Foundation in the U.S. estimates that the average sleep time has dropped by more than 20 per cent in the past century. By 1999, 64 per cent of people were getting by on less than eight hours' sleep a night and nearly a third were getting less than six hours. In Canada, a third of the population complains of "periodic" insomnia; up to half of these suffer it "chronically." A time survey of Canadian academics at five universities across the country not only found high rates of sleep deprivation and insomnia, it also logged high rates of memory problems (36 per cent) and problems concentrating (28 per cent). A significant minority (nearly 30 per cent) indicated that "I can't slow down enough to be in touch with my innermost thoughts."

Lack of sleep is not the whole story; it's also that we're not dreaming enough. The research is startling. Dating from the 1953 discovery of rapid-eye-movement (REM) sleep, scientists have courted the mysteries behind the dreaming phase of sleep. They've discovered that during "slow-wave" deep sleep (SWS), the brain's hippocampus is sending information to the cortex where it is layered into memory and memory associations. During the main dreaming phase, REM sleep, the cortex is active, sending impulses to the hippocampus in the evocative, metaphoric idiom of dreams. The result is crucial to learning and memory formation, say researchers. According to their findings, REM sleep building on SWS sleep not only makes us smarter, it knits the recent past into the fabric of the deep past, feeding the compost heap of memory association that tells us who we are, and even where we stand. Scientists refer to dreaming as a "sleeping-brain dialogue."

There's more. When we sleep in, drifting in and out of wakefulness in the luxury of nine or ten hours' rest, we're not only getting the rest some of our nearest primate relatives, the chimpanzees, suggest that humans need (although there's a fierce debate about whether seven, or even six, hours will really do), we are also tapping into a highly creative period. What scientists have dubbed "hypnagogic half-dreaming" is a phase of maximum creativity characterized by low-adrenalin arousal that allows for uninhibited, free-flowing thought. It signals that we are not only refreshed but deeply attuned to our selves and our sense of the world, capable of rich lateral thinking. Physicist Albert Einstein referred to this half-awake state as "combinatorial play" time and apparently indulged in it, even relied on it.

(Those who still consider sleep a self-indulgence should recall that dreaming was central to earlier generations and older civilizations. Dream time is where life comes from, according to myth and tradition around the world. For Canadian First Peoples such as the Dunne-za a successful hunt was experienced first in a dream. "It was only a matter of time before [the hunter's] trail on earth would catch up to the trail of his dream." It was a kind of foreknowledge that we often associate with intuition: the blending of background knowledge with fresh information through attentive perception. It is interpretation at a level that goes well beyond the efficiencies of rational logic and standardized categories of data. It is a precious part of our cultural heritage, yet currently, I fear, under threat.)

So much of what scientists have been telling us about the physiology of stress points to honouring the rhythms of life inside us and maintaining their integrity. Again and again, the research rediscovers the need for balance and equilibrium. It emphasizes the recuperative powers of our bodies and minds; when we give ourselves the benefit of sleep's inner dialogue, its rhythmic recursion can revitalize our sense of self and equilibrium. It's when we can't (or don't or won't) that our boundary conditions are breached and we begin to pay the larger, heavier price of stress and feeling overwhelmed. We lose our ability to perceive, accurately interpret and act on our reality. It's a reciprocal, mutually reinforcing downward spiral, I think, and dangerous.

TRASHING OUR SELVES

The lack of affect, the dulling and numbing of feeling that accompanies stress and burnout, signal that our inner equilibrium has already been breached. We lose our bearings, become uprooted and adrift from our sense of self, and then we can't even save ourselves. The hormones lubricating the immune system have run dry. Our inner vocal cords cannot scream a warning. The germs that are always present, waiting, are free to take over. The body is ripe for collapse, even as the hand on the mouse is still going and the other hand reaches for the tea or the super-jolt coffee.

To discuss this idea further I contacted Robert Romanyshyn, a California-based psychologist whose book *Technology as Symptom and Dream* analyzes how modernity is an escape from the body into the pure freedom of the mind and rational thought, using various technologies as an enabling medium. I was equally intrigued by his discussion of what happens to the body that has been left behind, neglected and abandoned. He's one of a rare few psychologists who consider the social context, the larger environment of psychological conditions. We e-mailed back and forth, then continued our conversation on the phone.

Why, I ask, does he describe technology as an "anti-aesthetic"— in other words, actively erasing our taste for truth and beauty.

Because, he replies, it separates us. "First of all, splitting that part of us that we say is mind from that part of us that is body, and then separating from nature. When we lose touch that way, when we put too much distance between our consciousness and our body, we lose what I call an aesthetic sensibility. The word means to "sense or to feel"—to allow ourselves to be touched by the nearness of things—whether it be another person, the morning light or the sound of a bird. It's the body that lingers, and that's inimical to a culture that wants to get on with things, that doesn't have time to stop and linger. So in a very curious way the question about time comes down to how technology is an anti-aesthetic. The fast-paced, speed-of-light existence is anti-nature, anti-body. It's an anti-aesthetic attitude."

In other words, I think, it actively drives us away from and alienates us from our ability to experience truth and beauty. I remind him that in an earlier conversation he'd said that if we lose our natal tie to the world and we loose the bonds of gravity, then we fly off into a state of pure light. Is that what we are doing? I wonder.

"Yes," he says. "I think the way we hold body and spirit, consciousness and nature together is through love. And if we become pure beings of light in this fast pace of technology, in this light speed we are living in, then we lose our animal nature. If that becomes dead, inanimate, I think we've lost the bond."

"So, consciousness implies a body," I venture, "a body in touch with other bodies, in relationships, attending to each other, with love, over time?"

"Yes," he replies. "How can you have time for embodied relationships if the kids are so stressed at school and then come home and the TV is in front of them, and Mom and Dad are so busy that you have to put fast food on the table or eat at McDonald's? There is no time for communal rituals, there is no time for conversation around the table.

"[If] we pretend that somehow we can distance ourselves from this natal bond of flesh, community and its connection to nature, then all of these feelings just go by the board and we become anaesthetic." He pauses, thinking, "There was a nice phrase that you used in one of your articles."

"We anaesthetize ourselves," I prompt. We discuss how it feels as if we're all patients anaesthetized on a surgeon's table but still with the ability to go to work and wander around shopping malls in this desensitized state.

I suggest that we're both the etherized patient and the attending physician administering the anaesthetic, and that this latter role is where the distinction between anti-aesthetic and anaesthetic applies. We experience the dislocation of our authentic selves living in real time and place, in real relationships with others, with shared memories, at the same time that we create these conditions for ourselves. We both take ourselves and are taken away from that larger, fuller sense of our humanity, considering values, priorities and meaning.

Dr. Romanyshyn agrees, "And I would make a distinction here. I think if you develop a real kind of feeling sense of your relationship as an embodied human being to another embodied human being, to nature, to the world… if you have that, then every present moment is a fullness and it draws you in and it has within each moment the fullness of a world, of an experience. The past is there, the present is there. Every present moment is horizoned by memory or imagination. The other kind of present moment that is so fast that it clicks in and out, it has no depth, it is like the flat image on the screen, like an electronic kind of being, it's here and then it's gone. It's the present as a quantitative moment rather than a qualitative moment of depth.

"I'm not anti-technology," he continues, "I simply try to read the inventions of technology symptomatically. So when I write about cyberspace I try to ask, 'What is it calling us to remember?'"

74 At the same time, he worries that cyberspace could become the ultimate split of mind from body and with it we'll experience terminal identity.

"Which is?" I ask.

"A terminal identity, for me at least, has two meanings. One is that we experience our identity at the end of a computer. We become related to the other only through distance, vision, through reading what is on the screen without all that aesthetic sensibility I talked about. But it could also mean that it is the end of our identity as human beings. The ultimate expression of becoming pure mind and discarding the body."

Romanyshyn sees hints of this terminal identity in two disparate diseases of our times: anorexia and chronic fatigue syndrome. He's written about anorexia as treating the body as an object to be controlled, even to the point of annihilation. And in our conversation he noted that chronic fatigue syndrome suggests that people have lost their sense of the natural, instinctual body and are living out of the image of the body as a machine.

"You mean, as if their body sense has been replaced by a sense of themselves as machine?" I ask.

"Yes," he says. "We can imagine our body as a machine and we can even try to live that way, but they are not machines and so the chronic fatigue syndrome is as if that part of us that knows better is saying, 'Don't forget you are not a machine, you are a part of nature, you have instincts and natural rhythms.'"

After our conversation, I sit back at my desk and consider all the gadgets that keep me going like a machine, driving myself to the point of burnout. I consider, too, how I'm on the ragged edge of chronic fatigue—and workaholism. I've played a dangerous game of chicken between them. And I'm not alone, I know.

Three

WORKAHOLICS
AND CHRONIC FATIGUE

*"Speed is the form of ecstasy the technical
revolution has bestowed on man."*
MILAN KUNDERA, Slowness

"We are the ghost in the machine."
CHRISTOPHER DEWDNEY, The Immaculate Perception

I MET ROB DUBÉ a week after he confessed to being
addicted to work. At thirty-three years old, with an
executive M.B.A. in his pocket, a wife and three
young boys at home and an upwardly mobile job in government management, he had the world by the tail. And
it had him.

Typically, Rob is up by 5:45 a.m., in the shower by
5:47 and out of the house before anyone else is up. "On
the way in to work, I'll grab a coffee because I need a
coffee to get me sparked. And I will check voice mail on
my phone in the car on the way in. I'll check e-mail. I'm
often driving with my left knee, scrolling on my Black-
Berry with my right hand and making calls with the
phone on my head. I'm often maximizing productivity
on that drive. So driving time is not downtime. In fact

it's extremely productive time for me. Closing communications, building relationships."

I jump in: "What do you mean, building relationships? Are you actually having a conversation with someone at, what, 6:30 in the morning?"

"By e-mail," he says.

"So you're building a relationship through e-mail."

"And through voice mail, yeah. And if I'm driving from this office to another. I have two offices because the second part of my job is technology. My job is psychoanalyzing a corporate client for their business needs and translating that need to a bunch of technical people so they understand what they're trying to implement. So I'll take advantage of commuting in my day." He breaks off, laughing, explaining that "it's a struggle trying to get through the linear process of trying to describe a typical day, because there are a gazillion tangents that are pulling me everywhere that I want to tell you about—and that's what a day is like, right? Just trying to get through that linear start-to-finish process."

I explain that understanding what he does in a day will give me a sense of him in context. And it seems that's precisely what he finds constraining.

"I'm an excellent improviser," he tells me. "I'm an exceptional improviser. I thrive in chaos. This environment (the federal government), they're in transition, corporate transition. And the crazier it gets, the more budget cutbacks there are, the more restrictive timelines are, the more unrealistic the workload, the *better* I can do. And in a totally sick way I love it."

"Why 'totally sick?'" I ask.

"Because you don't know when the end is. When you don't know what the end is, you don't know when to shut it off, and that's what's sick, right? You just keep on; it's like, hey man, I can do this, I can do that, we could do this, we could do that. That links with this, this links with that, let's do all that. If we do all that, then look how great we can be!"

His words strike a chord. Only, for me, it's more like appeasement. If I can just get this one report or chapter finished, I can relax. I'll be okay. Though if I still feel a little bounce in my brain, maybe I'll at least sort and delete the backlog of e-mails before I quit. What it means hardly matters, only that I've notched another tick on my to-do list. I tell Rob this, admitting that I guess there's a game aspect, too, a bit of a contest with myself to see if I can push myself just a little bit further.

"But with you," I say, "it's like you're always upping the ante."

"Totally," Rob replies, grinning. "I'm fifteen steps ahead. I'm not filling the funnel to its fullest. I'm building the funnel bigger." He laughs.

"Yet, it's almost like vapourware," I say, referring to software that exists more in the promotional promises of its developers than in a product itself. "It's all what you say you can do. The big game plan, the big numbers. Isn't it sort of like the fraudulent inflation of statistics that brought Enron down? It gets to be fabulist. Fictional. Because it's all just ideas—right? At the level where you operate?" I'm taking a chance here, knowing I could deeply offend him.

"Yeah!!" he says, still grinning. " But my staff have to deal with this, eh? I have colleagues and staff that have to deal with this."

He flips open his BlackBerry. Fingers flying over the miniature buttons, he pulls up his e-mail, gets into "edit" mode to reschedule his appointment with me from this week to next, then presses "send," knowing that a copy will automatically go to his secretary, who will deal with the chaos in my schedule and any resultant flak. I think of how angry and insulted I'd have been if he'd actually done that to me—especially so impersonally, behind the mask of an e-mail message I might or might not receive in time. I ask if he does this often.

"Absolutely. I'm extremely... " he pauses "irresponsible, I guess... but I don't mean that. I'm extremely loose and organic with continually shifting things that I just don't think are a priority. It's very easy for me to rationalize something not being a priority. Work is my priority. So I can rationalize anything not being a priority," including life at home with his family.

WORKAHOLISM AS ADDICTION

Rob's words remind me of psychologist Barbara Killinger's thoughtful book *Workaholics: The Respectable Addicts.* Workaholics, she argues, are classic narcissists: aloof from others, dissociated from their feelings. They become increasingly self-centred, preoccupied with what they must do and accomplish even at the price of personal relationships. They do this because they don't have a healthy sense of themselves. They're anxious all the time. They characteristically over-plan and over-control, with lists and lists of things to do, often broken down into detailed steps. They become single-minded. They develop tunnel vision. They compulsively drive themselves and increasingly cut themselves off from others—and even the capacity for compassion.

Workaholism is a vicious circle of addiction. The more time workaholics spend doing, the less they're able to simply be. The more they're thinking and planning, the less they're able to feel. Their powers of intuition atrophy. So does their ability to communicate with others. As they get more and more closed off in their work, crisis approaches in the form of claustrophobic episodes and anxiety attacks. They're more and more out of touch with themselves, less and less able to relax, to sleep, to stay asleep. Cross-addictions set in: to drugs, to drink, to jogging and exercise routines and to sex (as performance and new, often on-line, conquests). With an increased reliance on technology has come a host of new addictions: to shopping, on-line and off, and to on-line gambling, gaming and cybersex. The very act of going on-line can become a habit—cruising and schmoozing and seeing what's going on eBay. But the main addiction, albeit the least talked about one, continues to be work: the compulsive need to keep taking it on.

79

Very simply, "We live in a workaholic society," Killinger tells me when I visit her in her tranquil Toronto home to talk about this.

Her account is a fair reading of the Greek myth of Narcissus as retold by Marshall McLuhan. Narcissus, on seeing his reflection in a pool of water, mistakes it for another person. He gazes at that reflection, gesturing toward it and, receiving a response, gesturing back. The nymph Echo tries to win his love and attention by reintegrating him with his other senses. She tries to free Narcissus from the closed loop of his visual perception and gestures, by playing back to him fragments of his own speech. But he is numb and can relate only to that fragment of himself that is extended in the reflection, which holds him in a ceaseless playback of his own input.

McLuhan evoked the Narcissus myth to warn of a possible closure within the new environments we create through technological extensions of ourselves. "This extension of himself (Narcissus) by mirror numbed his perceptions until he became the servomechanism of his own extended or repeated image." A servomechanism is an automatic feedback-control switch in a closed system (the thermostat in a home-heating system being a simple example); McLuhan used the term to mean both the idea of people being drawn into a closed world of simulated reality and, once there, being subjected, however ineluctably, to its conditioning effects. The term allowed McLuhan to suggest the loss of personal will and genuine self-awareness as a human being.

It's equally chilling that McLuhan linked it to our increasingly wired social environment and further suggested that it's contributing to a "narcotic culture." (Narcissus comes from the Greek word *narcosis*, meaning "numbness.")

80   I SUMMARIZE McLuhan's argument to Rob Dubé, then I ask if he thinks today's environment of constant change aided and enabled with instant, asynchronous communication could foster workaholism, or at least incite its addictive, compulsive symptoms.

"For sure," he says. "It's like saying to a heroin addict, 'Hey, we've got these super, better syringes that can gradually leak heroin and that can hide under your shirt sleeves, and we're selling those for cheap, and you can subscribe for two years and, yeah, it's a ride.'" He laughs.

"I'm laughing because as I'm talking my BlackBerry's buzzing, and I came to meet you with my cellphone and my BlackBerry. What was I thinking? That I would have a phone call or be doing e-mail while we talked? But that's the whole thing of technology making things faster. You have the better means now to do what is your obsession, which is work! Fuck, you can wear these in your ear now: you have connections that are seamless. You have voice-to-text, text-to-voice. I'm looking at this, and I'm all over it. I can't wait 'til I have a headpiece with voice recognition that I can dictate to, that reads me my voice mail. Just watch what I'll be able to do!" He laughs even while he knows, and admits, that it's sick, sick, sick.

So where does it start? I make my own confession, telling Rob that I've struggled with workaholism most of my adult life. I, too, fit the classic childhood profile: raised by busy, hard-working parents, raised on a diet of conditional love ("good girl" for doing this or that, followed by a kiss, always in that order). Rob nods. His father was a senior corporate executive who travelled the world on business. His mother stayed at home but ran the house on the exacting standards of a good-housekeeping perfectionist.

The stuff of childhood might sow the seeds of workaholism, but often one event jolts it into bloom. With Rob, it was doing an executive M.B.A. while continuing to work full-time, as many people have done in recent years, retooling themselves after being "downsized," upgrading themselves as a hedge against more and more work being outsourced.

"In a lot of cases, the drive gets turned on in there," he tells me, getting wound up. "Holy shit, I *can* do it. I just confirmed something here. I *am* capable of sustaining this. My wife *can* handle this, you know! Do it for two years. Holy shit, I didn't know I could crank it up this high. Wicked. Now look who wants to hire these people. They can crank it."

I ask if he thinks that's where the needle gets installed.

"You're already an addict before you go," he says. "You just don't know it. That just confirms your tolerance, and your need. When I left that program, I had a much greater threshold, tolerance, appetite."

"To what?"

"To go out there," he says. "'Just bring it on, man. Just bring it on. Whatever!' It has nothing to do with the salary."

"Is the rush, the fix, just the doing?" I ask.

"Yeah, it is. The rush is doing."

I nod, and I talk about how hard it is for me to slow down. In fact, the more tired I get, the harder it is to quit. The insecurities set in. Just one more thing, I tell myself, and I can let myself off the hook. I can't break away, can't shut down. At that point, too, I tell Rob, I retreat behind the technology, the simplicity of expediting messages, filing stuff away for later, forwarding files. It's all just symbols, figments of my imagination as much as anything else. It's superficial and easy, yet quantifiable as work, as performance.

Rob adds another reason: "I don't want to deal with people sometimes."

"Why?" I ask.

"Because I'm too sunk and absorbed in what I'm trying to do, some state I'm trying to move toward. Too many things would take me away from that by communicating with those people in a live, face-to-face.... Too many variables. This isolates the variables in a theatre that is more conducive to completion of work."

"So it's sort of like tunnel vision," I say. It also sounds like the closed feedback loop of Narcissus and his reflection, more virtual reality than real.

Rob agrees. "And the end result, because it discourages that kind of communication, is a form of spiritual atrophy that you start developing. Everything becomes focussed on efficiency. That becomes the measurable standard by which one can rationalize one's existence from day to day. As opposed to the intangible, harder things in life."

"Like relating to someone else?" I venture.

"Yeah, like relating to someone else. Being part of someone's life."

I ask him if he's ever found it hard to be part of his kids' lives, even when he is at home and supposedly giving them his full attention.

"Oh yeah. For me to be with Jan—my second little guy, he loves to talk, he's so expressive—for me sit and listen to him talk about his day kills me. It kills me because it's not coming at me fast enough. And it's not intellectually challenging enough. I feel compelled to pay lip service. And I quite frankly think that my NLP is very easily read by any child."

"NLP?" I ask.

"Neuro-linguistic programming," he says. "My non-verbals, you know?"

"So basically what you're saying is that you're there but not really listening."

"I'm there and I'm not there."

"What do you mean?"

He shrugs. "I'll check e-mail. Pick stuff up."

I am shocked. "So, your kid's talking to you, and you're breaking eye contact?"

"Yeah. Absolutely. I'm very rarely engaged, looking. Very, very rarely."

I THINK of the stats on attention deficit disorder (ADD) among children and how they've ballooned since the early 1990s. It seems there's an epidemic of kids who can't sit still, can't relate, engage and stay focussed for any length of time. No wonder, I think. Researchers have identified as a factor harried parents for whom quality time with their children has dwindled to moments. I want to know why loving dads like Rob Dubé are treating their kids this way, even though they know it's bad. I want to better understand the addiction, particularly what's going on beneath the surface. I'm reminded of psychologist Bruce Alexander's research on alcoholism among the men who worked for the Hudson's Bay Company in the eighteenth- and nineteenth-century fur trade, many of them recruited from the Orkney Islands in Scotland. They were uprooted from close clan communities there and thrown into a life of constant mobility through strange and foreign terrain. They worked hard, canoeing deep into what, for them, was an uncharted wilderness, portaging supplies in and bales of fur out. And they drank. As Alexander summarized it in *The Roots of Addiction in Free Market Society*, alcoholism is a recurring feature of societies in the throes of rapid change involving "dislocation." Through drink these uprooted Orkney Islanders retreated into a world that they could understand and control, even as it controlled them in drinking's compulsive habits. They were no different, Alexander feels, from the drug addicts whom he studies and works with today on East Hastings Street in Vancouver.

"As painful as the drugs are, they are much less painful than the void," he tells me in a telephone conversation from his office at Simon Fraser University. The void being the state of having no place or stake in the larger society off the street. They're displaced people in a sense. "They have no real connection with anyone or anything. Imagine the horror of having no one, and then compare it with the relatively minor horror of being a junkie. To be a junkie is to be a somebody. People are afraid of you. They take all the trouble to arrest you and find social workers, and they want to come and interview you and give you a questionnaire...."

I ask him if the dislocation and uprootedness he studies implies time as well as space. He thinks they're inextricable. "To be cut off from time is, I guess, to be cut off from psycho-social integration, and I guess to be psycho-socially integrated is also to have a place in space. I belong here...." In its absence, addicts are often compulsively present-minded. We talk about some British research on addicts, which I'd sent him, that describes them as constantly improvising their lives in the present moment, their sense of time as continuity from past through to future having atrophied. He sees the same thing among addicts in Vancouver's Downtown Eastside.

"But it's also something you see at the university," he continues. "There are many pressures on academics, but many of us who are academics use academics addictively as well. We get caught up in this frenzy of delivering, and everything else falls by the wayside...."

How does the calm of focussed attention spiral into a frenzy of doing? I suspect it has less to do with the stuff of addictions and more to do with being uprooted: losing what has been a taken-for-granted rootedness in the rhythm and rituals of shared time and space. At some point perhaps we find surrogates in multi-tasking rituals and the frenzied rhythms of work.

THE EFFECTS OF ENTRAINMENT

In the seventeenth century, a Dutch physicist named Christian Huygens who was trying to perfect the pendulum clock noticed that the pendulums in all the clocks along his wall were swinging in perfect chorus-line unison. He didn't flatter himself that this synchronization derived from his having perfected the heft and hang of the pendulum. Instead he speculated, correctly as it turned out, that the clocks were coordinating with each other through vibrations transmitted through the wood along the wall. "This phenomenon, in which one regular cycle locks into another, is now called entrainment, or mode locking," author James Gleick wrote in his best-seller *Chaos*. Chaos theorists postulate that there is a source of order in the universe that is neither built nor regulated from outside, but emerges spontaneously from within its disorder and seeming chaos. Entrainment, as internally generated and largely involuntary synchronization, is an example.

Chronobiologists maintain that entrainment is how some of the fundamental rhythms of social life are imparted, at an almost physiological level, possibly even before birth. Entrainment regularly occurs among people living in close proximity to each other. Women who work or live together, for example, will often find that they get their monthly periods at the same time. It becomes more voluntary at the level of social interaction—for example, among members of a work bee or team—even contributing to what social psychologist Robert Levine calls the "unique temporal fingerprint" of cultures and subcultures. Whatever the connections, this capacity for shared tempo and rhythm is considered essential to social cohesion. Linguist and musician Ron Scollon calls it "the bond that enables us to move together."

Entrainment can work many ways. Just as people can become entrained with a group of co-workers through shared bodily rhythms, it's equally possible for people to become entrained to outside rhythms, such as a staccato voice on a loudspeaker or the entirely artificial pulse of a machine. If this pace is quickened, it can still be followed even to the point of the body breaking down, as Canadian documentary filmmaker Laura Sky discovered when she interviewed workers in a highly automated auto-assembly plant. One worker, wearing the telltale brace and bandage of someone with carpal tunnel syndrome, described the pressure to work quickly to meet the company's new "continuous improvement" performance standard, saying she used to dream about how she could speed up her actions applying nuts and bolts in her workstation. It's that much easier for this "mode locking" to occur if people are isolated from each other and lose the ballast of shared social rhythms and conversation.

Studying the rise of industrial-machine time and pacing in nineteenth-century Britain, time historian Nigel Thrift identified two important dimensions along which "retiming man's consciousness," as he puts it, took place. One of these was the hectoring discipline of fast-moving machines and "instrumental action" in general. But equally, he argues, it involved people getting used to "symbolic interaction"—in other words, to thinking in the pure idiom of ideas and the symbols representing them, as knowledge workers now do every day. It's hard to set limits for an intersecting, self-referential set of abstractions, especially when they're dynamic and all around you, not just at work but at home and in the car, the store, the gas station and even the washroom if you carry your gadgets with you. The quick-click pace can get under your skin, even into your blood.

IT SEEMS CLEAR to me that workaholics like Rob Dubé are being entrained by the pace; however, I also know that he's trying to break free of what he recently acknowledged as an addiction. I want to know who or what served as his nymph Echo, calling him back to the larger context of life and the people in it. After all, he has broken himself out of his narcissistic trance, has resolved to change, to switch off, to come home earlier than his usual 8:00 or 8:30 at night, so clearly someone or something interrupted the continuous feedback loop.

"What got through to you?" I ask.

"My son." Suddenly there are tears in his eyes. His face has gone slack, all his pumped-up energy has disappeared. When he continues, his voice has dropped in volume and pitch. "My son came to me with a piece of paper on which he'd written '7 o'clock. From Liam to Dad.' And that was combined with..., " he trails off and takes a big breath. This is hard for him, it's clear.

"Usually when I come home early enough the kids all come running. And I started noticing that Liam just wouldn't. He was just completely ambivalent, complacent. And I started realizing that part of his expectations are for me not to be there. It's built into his reality.

"Not all the fighting with my wife, or anyone else for that matter, could have moved me. But when you start looking at people who have no control, it's truly unfair."

I wipe the tears from my eyes and continue to ask questions, wanting the whole story. "But would all of them still be up when you'd get home, typically?"

"Yeah, they're generally all in bed. So they don't see me get up in the morning and they don't see me coming home at night. So I'm actually an absent father."

"Uh-huh," I say, aching inside. I ask about the note. "So, your son Liam just gave you a piece of paper on which he'd written '7 o'clock?'"

Yeah, Rob says, reaching behind him and pulling his wallet out of a hip pocket. "He wrote on a piece of paper and he folded it up and put it in my wallet. And it sticks in my butt. I can feel it in my butt. And I keep it there so I can feel it in my butt. So that I know it's there. So I know he's there." By now he's got the piece of 8 1/2-by-11-inch paper unfolded, and I can see the bold red-felt-pen message: "*7 Heures. Liam à Papa.*"

"Wow," I say. "And what is he saying there?"

"He's saying, can you at least come home before seven o'clock? Like, he's asking me not to come home at six o'clock, it's 'can you at least make it for seven?'

"When you realize that, you realize, like, I'm obviously out of control. Obviously I'm *way* out of control when my six-year-old son is telling me this. I can rationalize all that stuff with my wife— she's an adult that I can rationalize with. Or I can rationalize with a six-year-old and just mess with his head."

Later I get the full story from Rob's wife, Michelle, who talks to me from her headset phone while she tidies her kitchen. With three small children at home, she works only part-time as a self-employed accountant. But she works more than full-time as a mother and homemaker, she admits, making every meal from scratch and matching her mother-in-law curtain for curtain in the calibre of her home decor. On the night of the note, she says, she had an appointment to keep at 8:30 or nine o'clock. She'd reminded Rob periodically through the day that he had to be home in time. ("He needs to be babysat," she explains.) But even then he forgot. She had to cancel her appointment, and Liam heard her crying. That's when he wrote the note.

When Rob did get home, not much later, he found both Michelle and Liam in their rooms, crying. When he lay down with his son to comfort him, that's when Liam handed him the note.

I watch as Rob refolds his son's note and puts it in his wallet. He slips the wallet back into his pocket and looks up at me with a look not of relief and joy but of pain.

"God, I can't tell you how many times I would just love to go back and take the blue pill," he says, referring to a scene in the cult film *The Matrix* in which people live out their lives in a computer-controlled virtual reality.

"You mean going back to narcotic oblivion?" I ask.

"Yeah, ignorance," he confirms.

I'm incredulous. "You'd carry on being as compulsive?"

"And happy, right? Because I was completely ignorant and happy. I didn't know I had a problem. Now I've taken the red pill and I'm conscious.... So I have to live with that now." The red and blue pills feature in a pivotal scene during the film in which the hero, Neo, chooses to become fully conscious as an embodied human being and his mentor, Morpheus, greets his awakening with a famous line from French social theorist Jean Baudrillard: "Welcome to the desert of the real."

I'm tempted to rephrase Baudrillard and Morpheus a bit and tell Rob: "Welcome back to the reality you had deserted." Rob had neglected it to the point that his child had adjusted, and Dad was no longer part of his reality. As I shake Rob's hand and leave, I wonder: is it entrainment or addiction, or both?

SOMETHING ABOUT today's society has intensified the force and allure of both the conditioning environment and this form of self-conditioning. It's not only the unilateralism of private-sector values like instrumental efficiency over what had been a healthy mix of private-sector competitive individualism and more compassionate collectivity, it's how these values are embedded in the wired and wireless infrastructure underlying the lived environment of daily life. It's also how much this virtual space-time has displaced the original context of ongoing embodied relationships, which have traditionally sustained our social institutions and society as a whole. The effects ripple from personal self-forgetfulness and absent-mindedness to whole chunks of institutional knowledge and shared memory and commitment being lost. In turn, I think these losses help explain the crisis of accountability and public trust, which I'll delve into in Part III. Meanwhile, they've brought on this very personal crisis for many people, including for myself. I'm often hovering somewhere on the edge of some sort of crash, achieving a show of productivity but at the price of consciousness and even a genuine, rooted identity. In other words, I think that I am both conscious and simultaneously unconscious, out of touch with larger social realities and meanings. Once I've been "on" for a long time, I'm able to cope with only small, easily accomplished tasks, and I don't even bother to think what any of them mean, for me or for anybody else. I'm just focussed on keeping going, going places, doing things. Anything!

Even when I'm sitting back, supposedly relaxing, letting a lull in the conversation just be, I can feel the itch to say something, do something, get on with it. It's like water bubbling up to a boil; the prickling hot bubbles course through me. It's a lot like the desire to have just one drink, or one cigarette. The feeling consumes me, blanking out practically everything else. It's so strong and so real, too, like a wave heaving itself out of the ocean depths. I can feel it picking me up, especially if I'm standing in a slow lineup at a store.

It happens even when I'm supposedly relaxing in my Muskoka chair in the yard, or lying on the sofa. The music I'm listening to, the birdsong, the details of the breeze blowing through the maple leaves can't take me in, can't ground me in my body, through what my senses are absorbing.

My mind is buzzing, drowning everything out. It's ridiculous; I'm drafting what I will write or say in a voice-mail message to set up tomorrow's meeting. I might as well just get up and do it, I think, and before I know it I've flicked the switch. The computer is up and running, and so am I. I'll pay for it, I know. I won't be able to fall asleep, or I'll pop awake at 3 a.m. and lie there rehearsing, reviewing, the lists getting longer and longer. I resolve to play that much harder, an oxymoron if there ever was one. To unwind, I'll work out for an extra half hour. I grab my bag and head off to the gym. But once I'm there I count the strokes, mentally log each rotation of the exercise bike. The restorative effect won't come because I can't shake the pressure of productivity. Working out has slipped into the same old coinage.

It feels as if there's less and less of me in the picture—the me that can see with an engaged and probing eye, that can sense the underlying pattern and interpret what matters in the midst of all the data. It's not that I've been pushed out of the picture. I'm still here—my finger on the keyboard, my eye on the screen—but my attention is so scattered that my perception is trivialized. I'm being trivialized, too, by all the abbreviated messages I'm processing through my systems.

I read somewhere that the half-life of an innovation is getting shorter and shorter these days, and it occurs to me that this can be said about much of our experience. Anything we do—be it attending a meeting, preparing a report or sending and receiving e-mails, voice mails, faxes and letters—no matter how urgent and important it might be, is replaced by the next thing and the next, so fast that we only half-experience it. Then it's gone, and almost as quickly forgotten.

It's the unbearable lightness of being digital, off-line as well as on, of being virtually here, there and nowhere all at once.

CRASH AND CHRONIC FATIGUE

When I consider what this means to bodies that need to rest, to touch and be touched by others, I think of David Cronenberg's repulsive and brilliant film *Crash,* a highlight in his ongoing meditation on contemporary technological culture. It depicts people compulsively attracted by speed to the point of near necrophilia. They don't actually make love to dead bodies, though they screw in fast cars and at crash sites. They're obsessed with the danger, the compelling need to flip and fall back into a forced reconnecting with their bodies, because they're so numb, so uprooted in their fast-lane modern lives. They're effectively "deadened by the non-stop shock treatment of postmodern culture," and crashing, in a car accident, is their breakthrough back to reality.

The difference between Cronenberg's screen people and those suffering from chronic fatigue or, like me, at least many of its symptoms, is that the crash doesn't outwardly mangle the body. The mangling happens inside. There's a terrible pain in your joints, but no sign of outward injury. You can't stand up for long, can't get to sleep at night, but it's a neurological pain and therefore tough to quantify. However, it is not all in the mind, though there's a lingering, Darwinian bias toward that view. Perhaps that's why I was reluctant to ask my friend Jody Berland to talk to me about her chronic fatigue: because it suggests a personal weakness best kept private and close to home.

I HADN'T SEEN Jody for years. We first met in the late 1980s at a conference in Montreal, where this York University humanities professor had given what I thought was a scintillating paper. We became instant friends as we exchanged e-mails and corresponded avidly for a while, visiting when we could. When that petered out, I didn't think of her as having disappeared or dropped out; in fact I didn't give it much thought, because this is the pattern of professional friendships, especially these days.

I had to hide my shock therefore when I saw Jody at a conference in Toronto a few years ago. She was pale, painfully thin and hollow-eyed. The talk she was to give was the first thing she'd written in years, she told me in a whispery voice. She was nervous, too, she said, worried that she wouldn't get through it, wouldn't be able to stand at the podium for that long.

Chronic fatigue felled Jody in the late 1980s, just after she was involved in a car crash. She wasn't really hurt in the accident, but the shock of it was the last straw in a life of working flat out, teaching, writing, helping to launch and run a literary magazine plus, recently, having dealt with the stress of a marriage breakup. Her immune system broke down completely.

Now, visiting her two years after the Toronto conference, I can see she's still painfully thin and pale. She makes us some herbal tea, and we carry it upstairs to her work and resting space. Her beautiful white cat settles into her lap, and Jody starts talking.

"The reason I can't stand up is because part of this illness, whether it is chronic fatigue or fibromyalgia, is what is called orthostatic intolerance, which means you actually can't stand for long periods of time. It has to do with blood pressure." (In fact, orthostatic intolerance means that you can't be in one position for too long or your body wants to lock that way. The muscles tense up, becoming painful, and it is difficult to change position.)

"Oh," I reply, "I thought you just got tired really easily."

"Well, I do tire easily too, but this is specific," she says.

We compare notes from what we've read about chronic fatigue, including a good report on it, finally, in Canada's national news magazine, *Maclean's*. The syndrome seems to affect people who are overextended, women and men, and when it hits it "appears to carpet bomb the body's immune system, nervous and endocrine systems." It's as though the immune system, having been strained to the breaking point, turns on the body, making you almost "allergic to yourself," as Jody puts it. The symptoms include tender lymph nodes, muscle and joint pain, headaches, sore throat and soul-sapping fatigue that sleep does little to assuage. On top of that, the mind shuts down. Studies have chronicled short-term memory problems, loss of vigilance or concentration and attention-deficit problems, all contributing to a "general cognitive impairment." When Stephen Lewis, former Ontario New Democratic Party leader and Canadian ambassador to the United Nations and now active once again as UN special envoy for HIV/AIDS, came down with the syndrome, he told *Maclean's*, "To be so completely immobilized was the most depressing period of my adult life."

A good day for Jody means she can do some reading or writing, maybe go for a walk and do something socially for two or three hours at most. A bad day is when she has to go back to bed after breakfast.

"No matter how much motivation I have, or how much I want to do something, it doesn't mean I'm going to be able to do it. It's this kind of absolute powerlessness that's imposed, that's probably the worst of it."

"Do you get tired of being tired?" I ask.

"Yeah." She laughs. "If I have the energy."

"What do you mean?"

"Well," she says, "every form of conflict or healing takes energy. Even going to sleep takes energy. One of the things about people with chronic fatigue and fibromyalgia (an inflammation of connective tissue) is that [we] have major sleep disorders, and one of the reasons is that going to sleep takes energy—a certain kind of energy that [we] don't have. So I have been on major sleep medications for three years."

There's a third person in this conversation, or there appears to be later when I'm rereading the transcript. Her name is Gail Martin, an intelligent, highly educated person who's also been felled by chronic fatigue and fibromyalgia; she does occasional transcription work to make ends meet. I hadn't paid much attention to her, partly, I have to admit, because I didn't see her as a "professional" as I imagine myself to be. She was marginal, my interview-transcription assistant, with no more than chit-chat beyond that, until suddenly she popped up in the middle of the interview she was transcribing.

Note from Gail: she inserted in brackets. *Too true. One of the things that happens with fibromyalgia is that the muscles cannot relax. You are in a constantly tensed state. Your brain cannot relax enough to achieve a sufficient amount of* REM *sleep—many sufferers cat nap rather than sleep through the night. You wake up no more rested than when you went to bed—and often more stiff.*

I tell Jody that I often have difficulty falling asleep. That and staying asleep, not popping awake again, with lists scrolling through my mind.

"Watch it," she tells me. "That's not good." She strokes the cat on her lap, sips her tea. She continues, "I find it's like death. It's like I had a baby and no one wanted to know. Illness is kind of like that, especially a strange one like this, and that's okay in a way—I mean, I don't want to define myself solely in relation to my illness, although I'm complicit in that as well, but on the other hand I would like some help, and everyone wants to say you look great, you're fine, and then it feels like some sort of social faux pas to say 'Actually, I'm not fine'... It's hard, even in the community. You lose the support. Like, in Toronto, if you're not out and about they forget you. So I spent a couple of years here in complete isolation because, you know, if you're not going to lectures...."

"It sounds like being 'disappeared,'" I say.

"Yeah. There is this thing of me seeing everyone else keep going while I'm... well, I'm not totally not going. I'm still doing some [things], but there is a sense that I feel 'disappeared,' so that every time I go out into a social situation or a public situation there's a kind of anxiety that I never felt before in my life. Because

I'm not used to being with groups of people any more, and I have this anxiety because I think I've disappeared or that I'm behind, I'm out of sync. I don't know what's going on."

There's another pause while Jody struggles to bring out the words that will make me understand.

"Your brain can't process things," she tells me. "It can't see the connections between things." *(Note from Gail: This was one of the things that frustrated and frightened me the most.... I cannot coherently do something or sometimes even communicate, whether it be verbal or written. I just get so tired—I reach for information and it is not there and then I just want to cry.)*

I tell Jody it happens to me, too, sometimes. "When I'm really tired, the words on the page just lie there, inert. Even if they make sense as words, they don't move in my mind."

Jody's face lights up, yes! "They don't make any sense at all. The world sort of loses its light." *(Note from Gail: Ditto!!! What a neat interview—)*

Jody continues, "Sometimes when I am working, it's sort of like my brain starts out this big and then after an hour it shrinks and I'm working through this little peephole and if I can just remember how to get through this sentence.... There's not much left after two hours. I can feel my brain shrinking and shrinking and shrinking and I think if I can just get to the end of this sentence, the end of that idea, with this little tiny bit of mental focus that I have left, and everything else is kind of... it's like going blind, but it's like your brain is going blind."

"Sometimes when I'm writing," I tell her, "I have to go back to the beginning of the paragraph to pick up the momentum. I have to keep doing it, too, again and again. Even to revive my sense of where I am trying to take a particular sentence! I'm sort of—'Okay, come on, that last little inch.' I am aware of that—that crumbling, that falling away of the feeling for what I'm trying to say."

"And it's completely incapacitating," Jody says. "After that, I have to go to bed. Usually I can't, let's say, do some gardening or go for a bike ride, and that's what I try to do, to build up. If I can build up my physical strength then I will have more stamina and I'll be able to get to the end of the paragraph."

I ask if the disease has at least made her more aware, more attuned to her embodied self.

"No," she snaps. Then she reconsiders. "Possibly by just slowing me down. But the illness, basically it just makes me shrink. It takes me out of my capacity to be here.... All of my capacities shrink down so that my body is shrinking. One of the ways we talk about it is that you don't necessarily want to go more than a couple of blocks away from home because it is like you're small. You can't extend. You want a place that will shelter you because you are very small and you don't have the energy to extend yourself."

As I leave Jody's home, I reflect on my own situation: that I'm often able to extend myself much less than I let on. There's less and less of me really there, deeply involved, knowing what I want to say, making what I do count. Too often, I end up just going through the motions. Instead of stopping to question this, and to consider how dangerous my lack of engagement might be, I go along with it all. The superficial symbol sphere is even my front, letting me keep up appearances all over the place while not being fully present and accounted for anywhere! It allows me to be superficial; that's the nature of its abbreviated, simulated form of presence and involvement. I used to take the initiative on social issues. I used to organize meetings and participate. More and more, now, I fit the profile sketched by public-policy professor Robert Putnam in his book *Bowling Alone*, of citizens who participate virtually by making donations, signing electronic petitions, posting comments onto Internet list servs and chat groups, and who never attend a meeting. The computer has become a screen behind which to mask my crumbling capacity for genuine engagement, for getting out there and speaking my mind, listening to others and changing my mind, negotiating and possibly reconciling differences, coming to consensus and taking action.

If I can't do that any more, what good am I not just as a writer but as a citizen, a full participant in my world? If I'm spread so thin, if my attention is scattered over so many projects and people that I lose the strength of my convictions (and even a firm sense of what those are), who am I kidding? Why do I keep going through the virtual motions, processing more information, doing my bit to feed a doubling of the memory capacity of the machines every twelve or eighteen months, when my own memory lapses into brownouts more and more, when I can't attend to something long enough to say or do something meaningful?

I resolve to stay in touch with Rob and even more so with Jody, because we are fellow travellers, not just personally but generally. Moreover, if Jody's not part of the picture of living in this hyper-mediated new world order, if we let people like her become the digitally disappeared, the losers in a Darwinian struggle to adapt to the new environment, what hope is there of changing it? It could become a realm where only the most desensitized and alienated, only those with cockroach-like thick skins can thrive, and what kind of world is that for anyone, or anyone's children?

For me, too, Jody is a touchstone to the real world. Like the folded-up note digging into Rob Dubé's butt, she reminds me of the realities we could be abandoning in the rush and flash of digital connection.

Part TWO

INSTITUTIONS:

Living in an Attention-Deficit Culture

Four

VIRTUAL WORLDS AND
DESERTING THE REAL

*"Time-space compression always exacts its toll on our
capacity to grapple with the realities unfolding around us."*
DAVID HARVEY, The Condition of Postmodernity

*"An increasing array of life is processed rather than lived, recorded
rather than remembered and tracked rather than understood."*
DAVID ALTHEIDE, An Ecology of Communication

I'M BUCKLED IN for the ride once again, and this
time, oh my God, we crash!

But no worries! There's no blood on the as-
phalt, no glass in my skin. I don't even have to dust
myself off before I'm up and going again. I grasp the
steering bar and once more I'm airborne. I'm hang-
gliding over the Grand Canyon. In fact, it's only a flight
simulator in a corner of the Canada Aviation Museum in
Ottawa, and a fairly primitive one at that. Still, it draws
me in. For three glorious minutes I'm a bird, a plane, su-
perwoman, soaring between the walls of the canyon.
Quick, quick, I tell myself, bank hard. Quick, quick,   *101*
straighten out. The other way, quick! My heart pounds.
My body tilts, furiously trying to avoid the wall. NO!!!
OH no! Crash!

I've come to the museum on a whim, having read a lot about virtual reality, from the stunning technical know-how involved to love-hate tracts on its path to a "post-human" future. I have a vague sense that this experience of being transported virtually somewhere else and losing consciousness of the reality around me is germane to what has worried me into writing this book. Having tasted the virtual experience, I now realize that it does seem to invite us into a new way of ordering space and time. It allows us to be in Alice's Wonderland as if it's the here and now, though the "here" is simulated and the "now" fleeting. Equally, it offers the chance to be present and not present at the same time. That is, to be virtually present elsewhere, engaged, even embedded in other niches of existence, but only through the thin skin of symbols and the flickering fabrication of abstractions.

Looking back, I am amazed at how easily I was drawn into the illusion of flight. Once I'd placed my hands on the control bar and pushed my face between the blinders that shut out the real world, my imagination went along with the graphic images, at least provisionally. And then, with the momentum and instant feedback, my mind jumped right in there and filled the gap. I took on the specious flight of symbols as if they were real. In fact, I made them real, complete with convulsive jerks of my body and the rush of an adrenalin high. This is hardly post-human. Still, I've begun to understand that when space-time compression and speed-up come together in the quicksilver flow of bits on-line and off, the result is more than the sum of the parts. It can fundamentally shift our experience of reality. There are powerful trade-offs involved that we haven't necessarily considered as we entrust our powers of perception ever more fully to the screen. We gain a wider view but, at the same time, a more ephemeral one. We make real things happen through this virtual medium of simplified routines and simplified, standardized abstractions; yet we're so separated from the real-life consequences that it's hard to feel responsible if one of the symbols or one point-and-click action is wrong.

There is also a cost just to being alone behind a computer screen or a cellphone, being in touch with others not only from afar but asynchronously through text and voice messages. Too much absence from the realities of shared space and time, too much removal from fully experiencing them and it's that much easier to neglect them, even to abandon those realities altogether. Qualities that have to do with continuity and sustaining relationships, things like patience and fellow feeling, can atrophy or wither as the opportunities to exercise them are deferred or displaced. Even the ability to fall back into the pace that normal conversation requires can slip. The density of the real world gets to be too much, the attention and sustained concentration it requires too much of a drag. Or the foot of the mind gets stuck on the accelerator, and anything less than momentum and quick turnaround triggers anxiety, impatience and even rage.

I discovered something else in the flight simulator. I sensed a shift as I became attuned to the simulator's pace and friction-free way of being. That feeling, I imagine, could become addictive, offline as well as on. Having everything at my fingertips and done so quickly was so freeing, so easy, and gave me such a heady rush of power! My worrying about real-life consequences fell away because they didn't matter any more. They'd been deleted. That was part of the freedom, I'm sure, but also part of the ennui that set in after so many highs and just as immaterial crashes. After a while, playing the game became rather rote and boring; nothing much mattered any more.

What concerns me is not that we'll become hooked on the virtual world, though that is a possibility. What worries me more is that we'll simply get used to moving with this fast, symbol-leavened world, not just as individuals but as professional groups, as institutions and whole sectors of society. We'll start to regard this way of being as normal, and we'll begin to take our cues on what's real and what matters from instant symbol simulations. My worry is that our anaesthetization is becoming more collective and more cultural, because when we start accepting symbols as reality and operating on that plane routinely, we're not so much numbing our senses as we are creating a different world with a very different aesthetic sense. When we collectively allow quick data to become our reality, we lose our perspective on the big picture. And, because we don't engage with symbols in the same way we engage with people, we can lose the ability to relate directly to each other, to think critically about the data in shared dialogue.

As we allow statistics to stand in for the subtleties associated with quality and character, as a society our consensus on what we expect from social activities, including health care and education, can shift to reflect the operating environment in which these take place. We adjust to its superficial reality and provisional engagement, which is all that the data, the symbols and quick-click actions and reactions permit. In a sense we are creating our own virtual world, which is exciting and has many advantages; I recall how carefree I felt in the flight simulator because I was disconnected from real-life consequences. So I worry: If people are only tentatively, briefly and superficially engaged, how can they feel responsible beyond the moment they're involved in? What's to prevent any of us from thinking that it doesn't matter if we aren't meaningfully involved in decision-making? Yet while we're distracted by the data and deadlines of our virtual world, more and more real-world stuff is being left unfinished, undone or neglected, with possibly dangerous consequences.

I worry too about what happens if the symbols misrepresent the real world. What if the performance indicators don't reliably refer to things that are really central to performing, for instance, such services as healing the sick and teaching the young to think? What if the signifying chain (that is, between what's being represented as signs and what they mean in reality) breaks down? We've already seen examples of this in the accounting scandal that shook the stock market after the collapse of the giant Enron Corporation or, more urgently, in the fatalities that occurred in Walkerton, Ontario, when lethal E. coli leaked into the town's drinking water yet the expert in charge maintained that all was okay. What's to prevent public trust from withering away and, with it, engagement and commitment to all the real-life work associated with the public welfare? It's a real possibility, especially when people are stressed and overextended, or simply busy in their own zone of flows. Keeping the signifying chain going is central to many, many people's jobs and whole sectors of the economy these days. Yet our ability to discern and define what is real for ourselves, with our own senses and engaged in conversation with others, is being eroded and, with it, our confidence to challenge what we fear might be false or misleading.

I got some insight into this possible outcome during my time in the flight simulator. While my body thrashed around on the platform, responding to developments on the screen as fast as they came at me, I might easily have hit someone going by in the actual space of the museum. Yet I was totally unaware of this as even a possible danger because I was effectively not there, not present in my body, nor conscious of the realities around me. I could have done some damage and been totally oblivious to it. Ironically, it could have been the result of my acting responsibly inside the virtual world of flight that did the damage.

I don't really think virtual reality will arrive in everyday life any time soon, at least not in such obvious forms as my flight simulator. But that's precisely why I want to take time here to dramatize what I sense could be at stake as, through an accretion of small, seemingly inconsequential choices, adaptive strategies and a large dose of the unexpected, fabricated realities become a bigger and bigger part of the blended wired and wireless social environment.

Already, more than 60 per cent of Americans (roughly 45 million) play interactive computer games on a regular basis, and enthusiasts spend an average of six hours a day "in game." It's such a lucrative new business that producers are selling advertising spots inside these virtual environments, and they are planning to offer real goods through e-stores and franchises in these territories. It's a natural evolution, considering that the whole on-line environment is built and brought to us in easy monthly subscriber fees, and equally frequent upgrades, by big business. We no longer walk to visit a friend or get some exercise, we go on-line, even to work out. A hybrid exercise bike plus video-game PlayStation will allow players to steer their on-screen facsimile selves, called avatars, in a race while pumping their stationary wheels. eBay, "the world's on-line marketplace," is selling not only games but associated collectibles for real-world currency. One of the more popular on-line games, EverQuest, boasts nearly half a million player-"residents" in its virtual space and is estimated to have a "gross national product" of roughly $2,266 per capita, making it richer than many developing countries in the real world. There's even talk, among virtual denizens, of forming a breakaway republic!

Arguably the globalized version of capitalism facilitated by global digital networks, where electronic investments measure in the trillions of dollars each day, is a breakaway republic of its own. All virtual reality does is bring home how much the hypermedia environment has come of age, taking on the aura of an actual place with activities inside it more real than mere representation would suggest. Virtual reality highlights the shift that this environment makes possible, whereby the world of the symbol eclipses both nature and face-to-face relations in society. In fact, postmodern critic David Harvey argues that the whole focus of the capitalist economy has shifted, concentrating now on the production of signs, images and sign systems rather than on material commodities. Not only are ideas, signs and symbols its axis, but this post-industrial, postmodern economy runs on the speed and efficiencies made possible, and almost inevitable, by these immaterial abstractions. We can't see the borders and the structures that keep it in place the way we can see the screen and the blinders in a simulator, but the effects are the same. They can alter our consciousness and our fundamental sense of what's real. In that shift of attention, focus and engagement, they can also cause us to neglect what is outside it, in the living bodies of social institutions and society as a whole.

I'll get to this latter aspect toward the end of this chapter. But first I want to consider the more intractable topic of how our sense of what's real is changing, substantially enough that it changes how we act, not just as individuals but as groups and institutions. History affords a useful perspective.

MODERN, INDUSTRIAL-AGE CONSCIOUSNESS

It took me a while to figure out why I was so excited by *The Railway Journey*, a delightful book by German historian Wolfgang Schivelbusch, but it has finally dawned on me. What concerns me is not so much how reality is represented, but how an altered experience of reality can become not only public or shared but virtually irresistible, too. Schivelbusch's research gave me the perspective that I needed. He argues that trains delivered travellers into a new experience of living, even a new consciousness, that was as fundamentally different from the pre-modern as "being digital" is in the postmodern era enfolding us today. He quotes from Goethe's journal of 1797 to demonstrate how trains distanced people from what had been an intensely engaged experience of travelling in the stagecoach era. Whereas the pace of the horse-drawn carriage and the up-hill-and-down-dale unfolding of the route allowed Goethe to note not only cracks in the road surfacing but "[a] man climbing up a beech tree with a rope and iron cleats on his shoes," trains go so fast and cut their own path (generally straight, as opposed to following the contour lines of the natural world) that people no longer experience the landscape through which they are travelling.

Essentially, argues Schivelbusch, people experienced space, and time, in a new way. The rail line along which the iron horse ran was a key element. It was the product of a surveyor and railway engineer's plan designed to allow trains to move with maximum speed without derailment, through the vagaries of any geographic space. By combining locomotive speed and standardized tracks, the rail line and the trains it carried projected travellers along a new plane coordinated by linear time.

Writing with a similar sense of how much trains—and, with them, standardized time—changed people's world view, author Clark Blaise made reference to the Sherlock Holmes novels that were popular in the heyday of rail travel. For example, in *Silver Blaze*, published in 1890, Holmes entertains Dr. Watson while they're travelling by train from London to Brighton by calculating the train's exact velocity as he times its movement between the telegraph poles, which are spaced at precisely sixty-yard intervals along the track. Conversely, if he'd had some way of knowing the train's velocity but didn't have his pocket watch at hand, Holmes might well have used the telegraph poles to deduce the passage of time. The point is that Holmes was immersed in a self-contained, self-referential space-time environment (a prescient model of cyberspace), and he epitomized the ideal modern man in supreme command of his destiny and the contemporary milieu of rational efficiency that standardized space and time had brought into being. As for his heroin habit, well, I've talked about addictions already.

Not everyone in the real world was as fully (or compulsively) adapted as the fictional Holmes was to his new environment. In fact it was the problems and difficulties of adapting that signalled a profound social transformation was going on. Schivelbusch uncovered evidence of new diseases and complaints, such as "railway spine" and hysteria, which are analogous to carpal tunnel syndrome, chronic fatigue and depression today. Illnesses in their own right, these long-ago ailments doubled as signs of a wholesale adjustment or maladjustment to a new collective consciousness. "Smells and sounds, not to mention the synesthetic perceptions that were part of travel in Goethe's time, simply disappeared," Schivelbusch wrote. Yes, I thought as I reread that sentence. Train travel dramatized something that was more subtly unfolding as the modern age matured: a shift toward not just a perception but an experience of reality that was spread across ideas, books, photographs and panoramic views plus direct participation in the social and natural environment.

Because trains moved so quickly that travellers could no longer experience the countryside through which they were moving, the travel time became something of a void that needed to be filled. Magazines and light novels were peddled specifically as train reading and gained such widespread appeal that station booksellers, such as Routledge and W.H. Smith, operated lending libraries through which subscribers could drop off books at their destination. At that time, most people travelling by train either read, or closed their eyes and slept. (Today, we multi-task.)

Reading further, I discovered another prescient parallel. What those who looked out the window experienced was both shaped and framed by the train itself. Schivelbusch wrote: "The traveller saw the objects, landscape etc. *through* the apparatus which moved him through the world." (These days the window remains where it is—the computer on the desk, the laptop, cellphone or handheld—but the landscape is on the move.) But this wasn't the same reality that Goethe had experienced when the stagecoach was travelling so slowly along the pathways of regular social life that he could pick out the details of what the man climbing the tree was wearing. It was altered by the train's propulsion and, to a lesser extent, by the tracks projecting the traveller onto a plane that was removed from many of the realities passing by and became a welter of briefly noticed bits and flat panoramic images, rather than wholes that could be heard and smelled as well as seen. Schivelbusch described it as a "trivialization of perception" and a "dispersal of attention."

For all the reach and range of modern transportation and communication that was coming into being in the industrial age, something was also being lost. As Marshall McLuhan averred, there are psychic and social consequences when the experience of daily life becomes the almost split-personality act of being present and not present at the same time.

A POST-INDUSTRIAL, POSTMODERN CONSCIOUSNESS

Today, the railway lines are satellite beams and fibre-optic cables running behind the walls, under the floor and even, wirelessly, inside our pockets, and the journey they propel us on is not just for travel but for life. Wherever we are at any moment, our senses taking in our surroundings, we're also only a dial-up or a press-enter away from this symbol sphere engineered for instant presence. It's an enclosing hypermedia social environment that, for the one billion of the world's six billion citizens who have a phone, is inculcating a new social reality centred on fast, asynchronously flowing symbols.

Much that is the history of modernity set the stage, including the very notion of realities obtained through lenses and windows, or constructed from ideas, theories, texts, plans and schedules, then made real as people acted upon them by applying theories on development, following doctors' treatment plans and collecting photo-postcards. It's no coincidence that photographs became popular in the railway era, along with reading books and newspapers, and that these fabricated realities were accepted as though on a par with the realities outside the train window. But the point in the shift I'm trying to grasp here is that now realities aren't just constructed but generated. That's why Schivelbusch's account of train travel is so helpful—because it shows how the train pulled travellers into a different realm of consciousness and, with it, a different experience of reality. In a sense, the train generated this altered reality much as the hypermediated environment does today. And it affected people collectively, just as the wired social environment does now.

Lenses have taken over more and more of our seeing for us, both internally (through ultrasound and magnetic resonance imaging) and externally (through cameras, BlackBerries and high-

110

end cellphones). Communication went through a similar transformation; voice mail and keyboards have replaced our own speech and handwriting. Today's hypermedia environment offers a host of lenses and a whole range of vehicles, such as search engines and interfaces, and of journeying experiences, from sending "pics" or smiley faces to downloading music and modules of text or forwarding a text message. Relative to these, direct engagement and expression are mere moments of choice, including the choice to log on, to take on all these ready-made symbols and their often instant, logo-like meanings, and play along with them.

With so much perception and expression anchored outside our bodies these days, it's that much easier for our sense of reality to shift. In *The Age of Missing Information*, writer Bill McKibben argues that television immerses people in a different reality. As they watch television or just leave it on for six and more hours of the day, people internalize a reality-message that has less to do with the particulars of any program than with the non-stop flow of disconnected fragments, which are always upbeat, always present and always, no matter what, on to the next thing, usually with the hope for self-improvement through the latest gadget or product. Today, we live our lives as though the set is always on: We get so caught up in the momentum of ongoing action, turnover and change that we don't even notice how banal the recombinant symbols often are or how superficially we are engaged with the reality they offer on the screen and off, and how glib, almost prefabricated our responses sometimes are.

In many work settings today, people are expected to match data sets to data sets, indicators of their performance to "expected outcomes." Both sets of symbols are disconnected from the particulars of real people in lived time and place yet these become the reality people relate to, even if they're just down the street or across the hall from the real-life people involved. The disconnect can deepen, become normal even, because of the timelines involved. The nanosecond speed with which symbols can move, morph and be recombined into new patterns of daunting complexity leaves no pause in which these largely anonymous abstractions can be checked out for their relevance to us personally, or as professional teams or institutions. Yet what choice do we have as either individuals or isolated institutions but to adapt, or at the least to go along with it all, provisionally, tentatively, mastering the new language and grammar of the hypermedia environment? The scale, the pace and the changing complexity involved can leave us exhausted not just physically but psychologically and even spiritually. Momentum alone will keep many of us going, though often not knowing what's real any more or what, if anything, matters.

Like political theorist Ronald Deibert, I sense a compelling fit between descriptions of a postmodern consciousness as a patchwork of unconnected moments and experiences and being in a hypermedia environment. But I am uneasy when postmodernists calmly announce that the individual self associated with modernity is giving way to not just a disembodied but a "de-centred" self, an identity that is constantly reconstructed or even made up of multiple selves, in keeping with the many layers of reality on which we operate. As unsettled as I am by the concept of the de-centred self, I'm even more concerned about de-centred institutions, which, in losing their shared space and time, are also losing a shared and sustained sense of reality. And if all that's left are bits and pieces of contracts, work orders and performance-review audits, what difference will it make if institutions such as hospitals and universities are still designated public versus private? I want to check this out.

DESERTING THE REAL: CONVERSATIONS WITH
SOMER BRODRIBB, ARTHUR KROKER AND DAVID SUZUKI

Political theorist Somer Brodribb has written a scathing critique of postmodernism as a new way of understanding society and how it functions. Her book is called *Nothing Mat(t)ers*. Exactly, I thought when I read it. I picked up the phone and arranged to get together.

As we hang up our coats and set out our umbrellas to dry, I tell her about my experience of virtual reality and how I see it transforming everyday life and culture, and yet how postmodernists seem to simply regard this development as the latest change in the weather.

She laughs. "Well, now it's virtually nothing matters, isn't it?"

"Go on," I say.

"What we have are alibis. There is an increasing sense of displacement, lack of accountability through the play of masks on the Internet, the way the self is represented, or not represented. There are many monikers, many selves, that can be taken up, but they are alibis. Postmodernism itself, I think, is an alibi for the lack of accountability that is being materialized through the use of the Internet today."

"Why do you say 'alibi'?" I ask. I've looked up the word in the dictionary and it stems from the Latin word for "elsewhere" and reeks of covert culpability.

"Having alibis is part of the promise of virtuality: you can be somewhere and act somewhere else without taking responsibility for those actions, without being accountable for those actions, without being identifiable or recognizable...."

"Do you seriously think this could become how people live their lives?" I ask her. We're sitting across from each other in a grey-toned radio studio and recording our conversation for a CBC Radio *Ideas* program.

Her bracelet flashes as she gestures in response. "Yes! Dissociated, disintegrated, degenerated. It is going to require a new kind of analysis because we are used to talking about labour and history and embodiment, and these ways of analysis are being challenged by the disintegration, the disconnection that is being hyped—that this is a good thing, to be disconnected, to be dissociated, to play."

"By 'play,' you mean that ironically, right? Because you're also saying it's a lie, an illusion, a staged reality that can hide or deny the real thing," I suggest.

"Well," she says, "it is definitely a denial of responsibility." She pauses, then continues. "For example, you often lose track of the amount of time you might spend surfing the Web or trashing your e-mail. It is a process where you lose track of the body. You are not in contact with other bodies. The sense of embodiment is one where there is mutuality and coordination with others, continuity and cooperation. But that kind of awareness in embodiment is what is being displaced. You do not have to take the risk of embodiment or the mutuality of it when you are facing a screen.

"Of course there's a body sitting in front of that screen trashing its e-mail, and there is this illusion that you have this control, that you can reach through this space, moving at tremendous speed, making all kinds of connections. But that, in fact, is part of the facade. You may have many alibis but it is not a process of real connection."

Are we becoming less accountable, or even less recognizable, to ourselves? I think of myself zooming around on-line and off, dropping in on this conference and that on-line consultation, making token contributions. And I wonder, who's kidding whom?

I also wonder, what does this imply if it becomes normal for everyone, not just my temporary fallback (alibi?) when I'm stressed? If others are acting the same way, speciously present while they're really elsewhere, they're also assuming that someone else is fully present and taking responsibility for the larger whole. Or else they don't care. They won't worry about who might be left holding the bag, as long as it's not themselves.

AFTER MY CONVERSATION with Somer Brodribb I search out Arthur Kroker, co-author with Michael Weinstein of *Data Trash*, which depicts a "virtual class" of corporate executives and consultants exerting leviathan-like power and influence through their mastery and control of cyberspace over an underclass of dispossessed bit-part McJobbers, unemployed or unemployable. Kroker has built an international reputation both critiquing modern technology and championing its positive potential.

I ask him, "Is there perhaps a deeper dispossession taking place in which the whole fast-forward data-processing process turns everything and everybody into a sort of data trash?"

"Yes," he replies. For him it is a function of how much we as a society, as a culture, "have chosen to live within the media," not just with it. He continues: "Then in some ways your body or thoughts become trash itself. You become a speck of information in some ways, where information passes through you." In other words, I think, as McLuhan warned, a reversal can take place. We might embrace the new technologies as extensions of ourselves, but they can eventually amputate our traditional ways of being, in our bodies, and we end up acting more as extensions of the technology. As McLuhan put it in the last thing he wrote, we become merely "an item" in the data base, not only readily "forgotten" but "deeply resentful."

Kroker agrees, though he draws a distinction. "There are two kinds of data trash, two orders of data trash. There are the real casualties of the new economy—real human beings who get left behind because there's no way that they have any skills and there is no way that they can be fitted into the new economy. Literally they are left behind as the casualties, or the roadkill, on the way to the new economy.... "

"And what is the second kind?" I ask.

"I think in a qualitatively different way the winners [the corporate executives and technology consultants] are trash themselves, because that is the price of winning. They are turned into human remainder."

I consider for a moment what he's said and have an image of people pressing the "delete" button on themselves. I ask him if this is an appropriate analogy.

"Yes," he says. "Pressing the delete button is a good way of putting it. A lot of times people press the delete button on their own life and certainly on their relationships. That has really been brought on by the relentless acceleration and intensity and speed and the real competitiveness of contemporary culture." He pauses, then adds, "Something human has been lost in the coming-to-be of an advanced technology."

"What is that?" I ask.

"For one, a sense of time. An epochal sense of time as duration. Reflection on one's existence. An ability to really exist and have complex human relationships with others. We have already pressed the delete button on a number of things that are really indispensable to building up human culture.... "

Or, I thought as I remembered a conversation with scientist David Suzuki, we've pressed the channel-change button.

THROUGH PURE serendipity I spoke with Suzuki, the beloved host of CBC Television's *The Nature of Things*, about how we could be trashing not just ourselves but the living planet. I had gone to his office in Vancouver to show him a documentary I had produced about First Nations' approaches to science and technology, and he was lamenting the difficulty of putting any of the world's First People on television because by and large they haven't adapted to this medium (or don't think it's worth their while, at least in the fast-paced sound-and-sight-bite version of it that has become so standard).

"It's because they pause to think," he said, "and television can't stand silence."

Nor can television stand long sentences, let alone whole paragraphs of storied thought. Suzuki used to include interview segments of up to four minutes in length, but now thirty seconds is about tops. (In fact, the average sound bite is now seven seconds.) Today the show is a welter of sound and image bites, all geared to hold audiences not just through commercial breaks but in what he and others call "the competition for eyes." (All these fragments jostling for attention help foster restless eyes and equally restless

thumbs scanning the multi-channel universe at the end of their remote controls.) The effect has been a gradual disillusionment for
this environmentalist.

"Here I thought I'd be bringing people to nature," he says.

Instead, he tells me, he's brought them "nature hopped up on
steroids," an optimized, simulated version of nature. The CBC
doesn't send him to the Amazon when it's planning a program on
the jungle. Instead it sends a cameraman who sits in a blind for a
month or so, recording thousands of metres of film. Then, back in
the studio, an editor grabs the best images of a parrot, a jaguar or
whatever. Those symbols get combined and recombined, creating
a space-time compressed version of the Amazon. It's all sensation.

What's more, it's a misrepresentation of what the Amazon is really like. "The reality of nature in the Amazon is that nothing happens," Suzuki says. "Nothing dramatic. It's all subtle." It's the
infinite dance of life intersecting with life from the micro level to
the macro. "We don't allow the essential element of nature, which
is time. Nature needs time," he says. Nature is time, the living
breath of life weaving diverse organisms together in a vast interde-
pendent matrix. (Ironically, "matrix" now also means a computerized gridwork of graphics, though the original meaning of the
word is the cervix, the gridwork of bones that supports new life.)
To be brought back to nature's matrix requires time, too. You have
to be able to dwell in nature so that nature can come alive to you in
all its marvellous interconnecting rhythms.

What Suzuki is essentially telling me is that presenting nature
on television means neglecting its reality, deserting it, abandoning
it, even betraying it.

"Perhaps they should rename the show," I suggest. "Call it *The
De-nature of Things.*"

"That's cute," he says, but is clearly not amused. Even hurt, I
can tell, though he agrees.

"In a way, it *is* a de-natured nature that we deliver," he admits,
taking the conversation back to the camera crew returning from

117

the Amazon with weeks' worth of prime "footage," from which the very best, most sensational image-units are edited into an infotainment package. "And that's only the start of the problem. Viewers watching *The Nature of Things* then say, 'My God, I want to get down there as fast as I can.' More and more people want to go out and experience the 'real thing'."

"And the problem there," I suggest, "is that their idea of the 'real thing' is the living version of the images, which they want to consume in the same way, as an extreme image experience."

He agrees. "And that's what they demand, and that's what the tourist operators try to deliver, roaring from one place to another and basically editing out all the boring parts in between, which is where you would just sit and wait."

"And let themselves be drawn in, let themselves be moved," I add.

"People don't want to do that. They can't wait around for nature to reveal itself. They've paid a lot of bucks. They want to see the penguins. They want to see the seals. They want to see the whales. They want that jolt, and whoever delivers it has to be able to do so in their Zodiacs or their helicopters or whatever."

Nature is becoming just another "deliverable," with lethal consequences for the real thing. I recall reading an article about a popular tourist spot called Spirit Island in Jasper National Park in Alberta, and I mention it. "It's become such a branded destination, with so many people trying to get there," I say, "that it is dying."

We sit in silence for a while, contemplating the grain of the wood in the lovely boardroom table at which we're sitting and sharing a muffin.

"We love nature to death," he offers, quoting himself from his television special "A Planet for the Taking." Yes, I think to myself, we love nature abstracted from the warp and weave of life in which all living reality is embedded, including ourselves. And the more we experience it that way, as a symbol to be possessed and consumed, the more we can trample the real nature to death. Moreover, the same can be said for how we love and live life. We go with the flow of signs and symbols in our fabricated world, oblivious to the fact that in the real world of our bodies and the body of our living planet so much is being turned into a desert or dying—and too numb or dissociated to care.

I'm reminded of that line from *The Matrix*, which Rob Dubé referred to earlier. "Welcome to the desert of the real," Morpheus says to Neo in reference to the shabby remnant of the natural world left behind as the resources of the planet, humans included, are drawn into the cybernetic grid of the matrix, where the steak might taste great but it, like everything else, including identities, is fabricated from symbols that can be re-engineered into any other shape or identity, in an instant. The fully realized matrix of compressed space and time is all around us now—a fungible, volatile, flickering fabrication. Then, too, life on the ground has been so squeezed by the exigencies of cutbacks and downsizing that it can become barren and desert-like, by comparison with which the fast, facile realities of the symbol world become more attractive, at least in their easy accessibility and their ability to get the job, any job, done, now. Our perception of what counts, and even what's real, can shift to the point that what happens around us hardly matters any more.

The toll stress is taking on our bodies, our relationships and our social institutions, it hardly even registers. The qualitative stuff that is the pulse and breath of life itself can blur and be as good as gone.

But it's not gone yet. We can refuse to be lulled by the formulaic multiple choices of our simulated reality. We can become aware of the deeper trade-offs of deserting the real, especially in areas where people really matter, such as in health care.

Five

NURSES AND HEALTH CARE

"If quality of care is the goal of the health care system but
quality is neither measured nor evaluated, accountability,
just like efficiency, remains at the level of rhetoric."
JANICE STEIN, "The Cult of Efficiency"

"Money, money, money. You know what that means?
Move your bed in."
HANNAH MAHONEY, health-research administrator

"WHERE ARE YOU FROM?"
The voice seeps through the butter-yellow curtain around the bed next to where my friend is recovering from a hysterectomy. It's a nurse, talking to a new patient up from surgery. I hear more soft words, and more mumbles by way of reply, though they are gaining a little in volume. Finally there's a bit of a laugh.

Connection. It is the start of a nursing relationship, I think. And it occurs to me that the nurse probably did much the same with my friend, helping her find her way through what writer Susan Sontag has called "the kingdom of the sick," helping her confront whatever demons assailed her as the doctors cut through to her disease state, helping her summon the will—the "agency," as they say in the motivational literature—to heal or, if necessary, to come to terms with what cannot be made all better. I know it's happened to me. I remember once coming to after surgery with a hard and hurting truth lodged in the incision the doctor had made. I can remember not wanting to wake up, not wanting to regain consciousness. And I remember a nurse going to work on me with her words, with her eyes and with the caring touch of her hands mediating my obdurate silence.

The connection is not just the nurse soothing and exhorting, it's also the patient responding. When the relationship works, it works as two people paying attention to each other and entering a curiously intimate dialogue in which they are attuned or tuned in to each other. Theorists call it a "triadic" dialogue because it's focussed on a third party, which is the sick or broken body in the bed. The common goal of the dialogue is to move that body from sickness back toward health, from weakness back toward strength, from patient back to person.

Although it takes science and technology, timely diagnostics, surgeries and medications to heal injury, to cure and treat disease, to get someone back on his or her feet, complying with the medications and any therapies includes something much more subtle: a gently put question while a nurse positions the call button or a doctor checks the drip on the intravenous. This engaged empathy is so simple and taken for granted, however, that it doesn't get listed on the patient-care chart. It's also so all of a piece, so unquantifiable, that it can disappear completely in the computerized gridwork of evidence-based outcomes that verify that "quality" care is being delivered in the restructured, re-engineered just-in-time space of hospitals these days. In fact, it almost invariably, and ineluctably, does disappear there, on what has come to be considered the official accountability screen, because it's practically the antithesis of what can be computed. It emerges from the shared space-time of a human encounter, and it pretty well stays right there, because it cannot be separated from that physical state.

Combining skills and ethics, subjective and objective knowledge, nurses create a context of caring interpretation in which patients can begin to "make sense of" what has befallen them and, if it is possible, heal. It's part, too, of what current policy theorists now champion as the "virtuous circle" of a healthy health-care system involving healthy health-care workers working well with each other in healthy health-care workplaces. Like me and much of the Romanow Commission on Health Care, they suggest that Canada's public health-care system can best be renewed, with its core principles of equity, fairness and solidarity around good health for all if that circle is preserved. As an ethos, it's as old as public health care and the nurses, midwives and doctors associated with it.

THE ORIGINS OF PUBLIC HEALTH CARE

The idea of health care has a long history. Archaeologists have unearthed colossal statues of the earth goddess in the bodily form of a mother, often with multiple life-sustaining breasts. At a healing shrine at Titane on the Greek island of Sikyonia, the breasts of the earth-mother goddess Rhea were called Hygeia, meaning "health," and Panacea, meaning "all-healer." Hygeia and Panacea later became part of the pantheon of Greek gods as the daughters of Asclepius, the god of medicine, and were incorporated into the Hippocratic oath. Hygeia guarded health by prescribing self-discipline and a good environment, and Panacea used drugs and various hands-on actions to heal.

Public policies promoting hygiene and a healthful environment were important in ancient Egypt and were passed on to the Israelites and incorporated into the Mosaic code—the laws of Moses. They were also adopted by the Romans and inspired the appointment of medical officers of health, to promote public hygiene, as well as the first state hospitals to care for gladiators, soldiers and slaves.

Nursing in the Western world owes a lot to the healing attributed to Jesus of Nazareth and especially to the virtues of caring for one's fellow human beings as extolled in parables like "the Good Samaritan." (In a moving exegesis of this parable, priest-turned-philosopher Ivan Illich stresses the carnal act of the Samaritan being touched by, moved by, the plight of the battered and bleeding stranger. Responding with empathy and loving identification, the Samaritan then takes the stranger into his care. It wasn't a prescribed duty of ethnic kinship, Illich argues, but the engaged corollary of identifying with this afflicted human being.) First, Roman matrons practised the new Christian faith to which they'd converted by opening their homes to the sick and for other charitable works. Hostels and hospices for the care of pilgrims extended this caring tradition and laid the philosophical foundations of hospitals. (The words "hostel," "hospice" and "hospital" all come from the Latin *hospes* and *hospitem,* which mean both "host" and "guest.")

Nuns became the next nurses and have continued to perform this service throughout the Christian era, notably during the fourteenth-century scourge of the Black Plague in Europe. They also helped extend the moral culture of health care into the modern era even as it became more secular and scientific. For example, in New France three Augustinian sisters opened the first hospital in 1639, and the Ursuline sisters founded the first school of nursing that same year. More than two centuries later, the Grey Nuns of Ottawa set up a quarantine ward in their convent when a smallpox epidemic swept the city. They tended the sick and took the dead to the cemetery themselves.

On the "panacea" side of health care, bone-setting and bleeding were offered and increasingly, with the rise of modern science, stethoscopes and other instruments of observation gained stature along with surgical interventions, as barber-surgeons began to establish themselves in hospitals.

Until very recently, nursing has been regarded as among the last of the craft-based professions because its learning has always been rooted in practice and in apprenticeship plus formal instruction.

As such, it has carried on a tradition of knowledge based on experience, in dialogue and storytelling that predates the modern (scientific) objectification of observation into facts and data informed
by rational ideas, theories and technologies. And it's reflected in
traditional patient-charting and record-keeping, done by hand.
"When you're writing something by hand, it's more a continuation
of the care you're providing with your hands," Hanif Karim, a nurse
in a seniors' residence, told me when we were discussing the new
trend toward computerization.

Doctors have traditionally shared a lot of the same culture,
gaining experience through their internships, relying on patient
histories for interpreting symptoms in context and using their
hands to palpate the afflicted areas. In more recent years, however,
the growing number and power of specialized doctors and nurses
has tilted the focus of health care more toward fragmented tasks
informed by diagnostic tests, both performed in isolated moments
more or less separate from the context of ongoing patient-care
relationships.

Today, nurses are caught in a tug-of-war. On one side are those
who feel nursing should go with the flow and become more scientific. In other words, that it is important "to make visible and
legitimate the work that nurses do" through, for example, the
Nursing Interventions Classification, a project originated by a
group of researchers based in Iowa that quantified nursing work
into set tasks for which national and even international standards
(including training and credentials standards) could be developed.
On the other side are those who argue that nurses should defend
the legitimacy of their heritage in the caring human touch, even
championing the need for a humanistic counterbalance to an
increasingly technology-driven medical profession in order to
preserve caring alongside curing in a health-care system that continues to be a "moral enterprise." As one theorist put it, helping a
person move from sickness toward health "involves the kind of
knowledge of the person that comes through an attentive gaze and

heartfelt listening." It's a reading of "responsibility to the patient" in health care that holds true to the word's roots in the Latin verb *respondere*, "to pledge, or to respond."

Striving for a middle ground, the one-time head of the Canadian Nurses Association, Ginette Lemire Rodger, actually quantified the effects of this implicated nurse-patient relationship so that she could get the subject of relationships onto the agenda of health-care reform and restructuring. In a daunting piece of Ph.D. research, she attempted to measure how empathetic nurse-patient relations can demonstrably reduce patient stress to promote the kind of inner equilibrium that is considered crucial to healing. Using a battery of technologies including an electromyograph (which measures the electrical activity of muscles and skin), an electrodermal graph (which records skin temperature) and dinamap (which monitors blood pressure and cardiovascular changes), she catalogued patients' bodily responses while being attended to by the kind of nurse who is capable of, as Lemire Rodger put it, focussing her attention on the patient and "being all there." Predictably, the signs on the graphs all told one story: the patient relaxed, which is considered crucial to letting a nurse get close, cultivating trust and motivating the patient to increase participation in her or his own care. In other words, patients "are inspired to want to care about themselves."

Lemire Rodger concluded, in 1995, that this type of nurse-patient encounter (which she called the "Z" interaction) "is powerful enough to create effects that are discernable in spite of the extraneous variables of clinical research with a very sick population" (in the case of her research, cancer patients undergoing chemotherapy) and proved the case for meaningful nurse-patient relationships in effective health care. (Further supporting this finding, a cancer clinic in British Columbia complements radiation, surgery and chemotherapy with massage, meditation and enormous emotional support to patients as agents of their own health and recovery. Coincidentally or not, B.C. has the highest cancer-survival rate in the country.) However, this whole discussion seems to have been supplanted by an entirely different set of arguments. In the name of cost containment, efficiency and, more recently, evidence-based accountability, health-care "delivery" has been fundamentally restructured and so, along with it, have the expectations of nursing work.

THE COMMODIFICATION OF HEALTH CARE

Concerns about the cost of public health care in Canada gained official momentum in the mid-1980s when Prime Minister Brian Mulroney's Conservative government launched a regime of public-sector spending cuts in the name of deficit reduction and fiscal responsibility. Because nurses represented the largest single labour cost, especially in hospitals, they came under extra scrutiny. There were some layoffs, but largely the number of nurses per capita declined in most provinces through early retirements and by replacing full-time positions with part-time and casual nurses hired on an as-needed basis. At the same time, other health-care costs were rising, as some of the big players in corporate technology—including McDonnell Douglas in aerospace, IBM in business-information systems, plus the big pharmaceutical companies—installed computerized information systems everywhere from medical diagnostics and record-keeping to medications dispensaries. Part of their success in winning big contracts at a time of health-care cutbacks stemmed from a promised agenda of bringing "out of control" and "unmanageable" health-care costs into line through hospital information-management systems that would keep track of everything.

Words such as "unmanageable health-care costs" were deployed like a rather large fishing net, in that they were applied to any costs that had not been spelled out ahead of time and brought under centralized management control. It didn't necessarily mean that the costs were out of control, though certainly that was the implication. These were costs associated with initiatives taken by

126

doctors and nurses in the moment, on the ward or wherever, as they exercised their personal and professional discretion, often without a paper trail of quantifiable accountability for each action taken. They were costs managed independently and sometimes collegially by work units, such as nursing units, within hospitals and other health-care institutions, not by hospital-administration authorities.

Traditionally, head nurses decided the care needs on any given ward based on their assessment of how sick the various patients were and how great their need of attention. It didn't matter if they spent a few dollars more on some days; the cost was not their primary focus of concern, though it was assumed that they'd be mindful of the money and of the citizens keeping watch through the hospital board. Like the head doctor, the medical director who traditionally ran the hospital, head nurses directly determined what quality nursing care should consist of. They were accountable for ensuring that this standard was met in the health and well-being of the patients on their wards.

With restructuring, practically every aspect of running a hospital—from ordering supplies to bedside patient care—has been reorganized around the integrated health-information management systems. Under the new system, head nurses are now called nursing-unit administrators. They're part of the hospital-management team and they are accountable to its information and accounting practices. The effect is two levels of reality: as before, the reality of caring for patients in the ward, hour by hour, day by day, plus the new data-based reality of patient-care outcomes and related patient-care orders, diagnostic reference groups and timelines. For members of the management team who do not work directly with patients, this second, data-driven reality flows to them via hospital and related health-information networks and it becomes essentially the only reality they ever see. Moreover, this standardized patient data is only one part of a larger data-based reality (which includes, for example, regional, provincial and even

127

national health-care delivery plans and schedules, ideas and standard benchmarks) that the hospital administrators now work with on a daily basis.

Nursing-unit administrators, along with other members of the hospital-management team, are increasingly being drawn into the second reality as they try to run their end of patient care as best they can in an environment of trimmed budgets, rising absenteeism and unexpected outbreaks such as flu and SARS, at the same time that they try to balance their rankings against standardized benchmarks for performance in the "public market" for health care to maintain crucial government funding.

Equally, being a nurse these days means navigating between the virtual symbol and keyword maps that quantify patients and the here and now of actual patients in the wards where they're assigned to work. Two documents, a "patient-assessment form" plus a daily "nursing worksheet," are the link connecting the two. On these, nurses record what they've done, what medications they've given and any changes in the patient's health during their shifts. The nursing-unit administrator takes that first-hand information from these forms and interprets it through a grid of standardized categories of care, each of which contains several corresponding levels of care. The result is a new document called the patient-classification form.

Time-motion studies and other techniques have quantified the time-unit required to complete each patient-care task (complete with standardized variations for gender, age and other demographics), and this information has been worked out for each category of care and each level within it. So, once the nursing-unit administrator has filled in the patient-classification form it determines the "timed-care requirements" per ward per shift. The next step is to translate this information into the most efficient use of nursing-labour time. First, the data from the patient-classification form are fed into a computer, where software massages the information against predetermined "workload indexes" with related assumptions about nursing productivity to determine the "care time" to be assigned for the next day's shift. As a final step, the management-team members responsible for day-to-day staffing allocate these care tasks among full-time and part-time nurses, plus practical-nursing assistants.

As they come on shift, nurses are expected to square the prescribed "care-intervention list" with their patients' reality through an additional document called a "patient-care plan," on which they mark off the completion of the standard care tasks. This leaves a data trail representing patient-care tasks duly ticked off that can be matched to documented patient-care needs. It concludes an information-feedback loop that in turn is ready for an accountability audit, demonstrating that quality patient care has been delivered in a timely manner. From a nurse's point of view, it also concludes a massive transformation of health care. In the words of Doris Grinspun, executive director of the Registered Nurses Association of Ontario, it "shifts the attention away from the patient as a person, as a whole, and places that attention onto a series of tasks to be successfully completed in the most efficient way. It is the antithesis of patient-centred care."

This shift isn't total, and it doesn't follow automatically from the new document-based way of running health care. It's the result of a combination of many factors, including a move toward just-in-time staffing, which means that more and more nurses are being hired on a part-time and casual basis so they can be assigned only as required to meet exact staffing needs in each ward for each shift. Altogether, nursing time is so tightly scripted to the expected workload requirements that there's little discretionary time left for nurses to deal with the unexpected, just to be spontaneous or to take extra time with a patient who is lonely or distressed. It's no surprise to me, then, that nurses are suffering more burnout, fatigue and stress-related ailments than almost any other profession. They're so stressed and scattered these days that job satisfaction has almost eclipsed pay and job security as nurses' primary concern: Nearly one-third of Canadian nurses are actively dissatisfied with their jobs—a rate more than three times the average for all workers. Not surprisingly, absenteeism and sick leave are up as well.

THE HUMAN COSTS

I asked some Toronto-area nurses about their experience of working in hospitals these days, and whether they feel they can still adequately attend to the patients in their care. Yes and no, they told me.

"I remember when I was beginning to get snappy with patients," one of them says. "That's absolutely not me. I felt like I was the only one left who was being kind to the patients, and I was proud of it. But then I started snapping. You know, being short instead of caring and warm. You can see their reaction, the body language. The relationship sort of folds."

Another nurse continues, "You lose your tolerance for other people's stressed behaviour."

"That's a nice way of putting it," the first one says.

"You mean the patients?" I ask.

"Anybody," she says. "Anybody who comes in. Most people in a hospital, when they come in they are stressed and worried and probably not themselves. And normally you have the ability to be there and help them. They're not always perfect patients, but that's okay. But when you are really busy and you have too much to do, it becomes impossible to handle other people's stress."

"And it's not just the patients," another nurse adds. "It's the physicians who are trying to grab beds and OR [operating room] time, and of course their livelihood depends on getting their stuff done. They are sometimes discharging patients too early because they want beds, so there's this huge tension going on around proper times to discharge people."

Everyone sighs.

She continues: "Half the world has terrible family dynamics that really impact on their health, and that is at least as important as why they came in, and why they are returning [to the hospital]. But you just don't have time to get to know the person and know if they're taking their meds correctly—because half the world isn't taking their medication correctly. You get so that someone comes in and it's always tasks, tasks, tasks. It stops making sense."

Nurses' assigned "core-care hours" are supposed to fill only 45 per cent of their working time, leaving them ample discretionary time to be with their patients plus plan, consult with others and document their activities. Yet with fewer nurses all around because of cutbacks, and more critically ill people now occupying hospital beds made scarce because of closures, nurses find they're run off their feet. Clearly, the faster pace of work plus all the fragmenting, the disembodiment of action into standardized tasks and reporting data several times removed from a relationship with a person is having a desensitizing, even alienating, effect on nurses. If doing something for a patient "stops making sense" in terms of a healing relationship between the nurse and the patient, it doesn't matter so much how a task gets done—or if it gets done meaningfully at all—as long as it's ticked off the list.

I ask the nurses about their experience of multi-tasking. One told me categorically that nursing isn't a list of fragmented tasks. "But it's becoming that," she says. "The nurses have to get this done and this done, and you can't possibly do what you have to do. It isn't a time-management aspect. People are so much sicker than before. They are tremendously sick. So it's 'this patient can't go another day without getting the dressing done,' etc. It does turn into a list of tasks."

I can sense an iterative effect at work. Traditionally, nurses spent time getting to know each patient, while giving them a bed 131 bath, for example, and finding out who's at home, who's reliable and whether there is any mental confusion or memory problem that might make medication compliance a problem. Not only are these more leisurely, easy tasks now assigned to lower-paid nursing assistants, time-pressured registered nurses (RNs) almost embrace the shortcut of the predefined tasks to be done—and, almost fittingly, a lot of these are technological procedures such as replacing an intravenous bag, measuring the blood-oxygen levels for the patient on oxygen, noting the heart-monitor readings—because these can be completed quickly and easily and ticked off the list. Between the go-go-go pressure of the tasks and the attenuation of contact and engagement with patients, nurses' job satisfaction goes down. The result, research has found, is that nurses withdraw emotionally.

Overextended full-time nurses are also coping by passing off shifts to an ever-growing pool of part-time and casual nurses. (Traditionally, about 30 per cent of nurses worked part-time while raising a family; by 1998, at the height of the restructuring and cutbacks, an all-time high of 48 per cent of nurses worked part-time. In 2001, 56 per cent of nurses under the age of thirty-five were working part-time even though most wanted full-time hours and often took shifts at two or more hospitals.) Although this helps individual nurses carry on, it further fragments nursing care, adding yet another strange face at the bedside and another gap in the continuity of caregiving on the ward, which erodes the shared knowledge that develops among people who work together over time.

Research has found that the casual and part-time nurses are also less involved in the goings-on of the hospital as a whole. They're not only less "in the loop" of the institution's culture and politics, they're scarcely present at that day-to-day lived level of reality. Research into the "psychological contract" side of health care shows that for part-time and contract nurses the work is seen as more "transactional" than a real commitment by both parties.

So what happens when nurses try to redress the imbalance they feel in their work with patients, when they choose the person

132

over the task list and the paperwork? One nurse who works in the delivery room explains, "I know I have a bit of a reputation for being slow compared to the others. And it is because I, for the most part, have absolutely refused to get to that stage. I go at the pace that the patient needs to go at, and this means establishing that contact and interpreting their body language and helping them identify what they need. I take perhaps an extra forty-five minutes to get through the prep and the birth and the tidying up after than anyone else. That is a considerable chunk of time to most people. There are a few around who recognize what I am doing and don't say anything, and there are others who are quite resentful and comments are made, that 'she is just in there with her patient.' Kind of like 'she doesn't want to come out and help us with the work.' It leads to isolation."

And yet effective teamwork has been shown to affect "health outcomes" positively, probably for much the same reasons as continuity of caregiver enhances continuity of care. As people get to know and trust each other, they open up more, they confide more, consult and collaborate. It's not just the emotive support. It's all the lateral communication; for example, passing on the news that the cleaner on the floor speaks Urdu and can translate for the patient in Bed 2. I think of my own experience, spending six hours in Emergency at the Queensway-Carleton Hospital in Ottawa with my mother, who'd just broken her hip. The place was crowded, every seat in the waiting room taken, the ambulance gurneys lined up in the hall. Through it all, the triage nurse wove her way around the room sorting things out, moving things along. Everyone from doctors to nurses, nursing assistants and orderlies stayed in tune with each other, making eye contact at key panic-edged moments, someone moving to someone else's side to touch a shoulder, to help. The place remained calm—well, relatively. This, in Emerg! I was amazed, and impressed. It was the incredibly synchronized eye contact, I concluded, and the trust and mutual attunement in shared time and space that this bespoke.

In the ward, though, nurses can sometimes feel isolated even when they're with the patients, because of all the machines in the middle. As one nurse saw it: "There's more and more technology, brain surgery, heart surgery, emergency, invasive and complicated procedures, and you get so that the technology takes as much time as the patient, and even more. If you have everything attached to the patient and you don't have time to talk to the patient, you just look at the machines." The patient behind the curtain can become a set of numbers and health indicators matched to a standardized diagnostic category: the hip in Bed 1 or 2.

A pattern of dissociation begins to emerge. It's only a pattern, but it might shed light on the startling news that between 9,250 and possibly as many as 23,750 patients who died while in Canadian hospitals in 2000 did so because of some preventable mistake in their treatment before, during or after surgery, or even without surgery being involved. One of the study's co-authors, Ross Baker of the University of Toronto, explained, "It is usually a series of things, and at the same time a failure for others to notice, either because they are too busy or distracted."

Doctors and nurses still do take time to truly be there for their patients, but they also know that what's relevant in the feedback loop of official accountability is what can be counted: bed turnover rates (keep them high), hospital readmission rates (keep them low) and health-care indicators. For most nurses, as long as the patient-care tasks are ticked off quality health care is seen to have been delivered. Not surprisingly, the trend now is that nurses no longer fill out their daily charts based on their observations and impressions of each particular patient in their care (they're not kept as part of patient records anyway), they just tick off each square of the standard-care indicators on the chart and write in something about an individual patient only if that person's needs clearly deviate from the standard care and care levels designated for his or her diagnostic reference group. "Oh, you have to be pretty sick for us to write anything about you now," one nurse told a team of health-care researchers. So who is focussed on the reality of the patient? I wonder. Who is ultimately responsible for looking at the big picture and ensuring that health care remains account-able to human realities?

RETHINKING ACCOUNTABILITY

When political science professor Janice Stein's eighty-five-year-old mother was hospitalized with a broken hip, Stein experienced first-hand how misguided our notion of accountability seems to have become. Seven days after her mother was admitted, while Stein and her sister were still searching for a new place for their mother to live, a frantic "discharge coordinator" approached her, wanting the old woman gone. It was nothing personal; in fact, it was totally impersonal. "Your mother is now a negative statistic for this unit," the coordinator told her. "Every additional day that she remains in hospital, she drives our efficiency ratings down." Having never imagined her mother being a "negative statistic" in a standardized efficiency rating, Stein was insulted, but she rational-ized that it was the system that forced the coordinator to speak and act as she did. Stein used the experience to write "The Cult of Efficiency," her Massey Lecture on CBC Radio challenging the fetishizing of efficiency and what it's doing to the integrity of accountability, particularly in crucial public services such as edu-cation and health care.

Essentially, she argues, we have shifted from a society sustained in health and well-being (to a degree, at least) by a universally accessible health-care system to a society focussed on indivi-dual health services delivered with maximum efficiency. Public-spending cuts, government-business partnerships and the adoption of businesslike practices and language have revolutionized health care and education, and efficiency was the rallying cry to do this. Not only was it used to justify "cost containment" and related control but also "competitiveness," and the public bought into this shift toward a more businesslike approach because it came in the

135

language of convenience and more client-focussed service, which appealed to us as consumers.

Stein drew on writer Michael Ignatieff's work on "the rights revolution," tracing how this evolved from an initial focus on human rights, equality rights and other collective rights in the 1960s and '70s to, now, the right of radical individualists to have maximum free choice in public and private markets. She linked this revolution to parents' demands for choice over the schools in which to enrol their children and to the public's demand for the latest diagnostics and treatments in health care, and timely access to them.

At the same time, an increasingly material rendering of individual rights went a long way in enlisting support for a radical transformation of the public's conception of services—from public goods indivisible from the common welfare to private goods on a par with commercial goods and services traditionally supplied by business. As a corollary, the internal makeover that occurred as business practices were applied to these core public services was accepted almost without question, with both markets and consumer choice assumed to be efficient. However, it was at this point that some troubling contradictions began to appear. Public services began to be defined almost entirely in self-referential terms of efficiency. That is, instead of "efficiency" referring to the human and social values informing the larger context of these public services (so the *effectiveness* of all the new efficiency techniques can be validated and affirmed), efficiency now has data referring to benchmarks and standards of efficiency and performance only. They don't refer to a human reality that can be checked out in a dialogue about actual experience. Instead, it's become efficiency for the sake of efficiency; cost containment for the sake of cost containment. In short, Stein argues, a cult.

To be meaningful for us as a society, Stein insists, public accountability must refer "results" and "evidence" to social values and societal purposes and, as such, it should be democratically ne-

136

gotiated. In fact, she goes on, "Much of the democratic debate of the next decade will turn on whether accountability is imposed or negotiated."

Public concern for "transparency" and "accountability" is growing, or it seems to be, given how prominent these words have become in politicians' rhetoric. The danger, though, is that the public lacks the time and opportunity to really be involved in examining what these terms mean, and so do politicians. And so, while politicians continue to use the terms, technocratic experts are proceeding as if no debate is necessary. As an example, the Romanow commission on the Future of Health Care in Canada, a government commission headed by Roy Romanow and established "to review Canada's health care system, engage Canadians in a national dialogue on its future and make recommendations to enhance the system's quality and sustainability," heard from a lot of Canadians, Canadian organizations and non-profit groups who argued the case both for efficiency and for a renewal of an ethic of caring. Romanow's report struggled to reflect this balance. But though the title and the proposed health covenant talked about values, the recommendations for institutional action emphasized statistics and "evidence-based decision-making."

The doctors, nurses and nursing assistants who work within the system are already expressing considerable skepticism, even resistance. They can see the benefit of standardized categories of data and classification systems for useful comparability among institutions and across jurisdictions, but they also cherish their autonomy and the discretionary space-time in which to exercise it. They're used to being trusted to take responsible action in the context of the here and now, and they want that kind of accountability to be honoured. Romanow's priority of a new National Health Council to "act as an effective and impartial mechanism for the collection and analysis of data on the performance of the health care system," with a focus on "accelerating the establishment of common indicators and measuring the performance of the health-care system" doesn't sound as if it includes much room for this kind of decision-making or more qualitative "evidence."

This evidence-based decision-making, and attendant standard health indicators, are central to devising an integrated health-care information system that can seamlessly include not only hospital-based care but home and community care, plus prescription management and drug dispensing at pharmacies, possibly for a national Pharmacare plan. Yet, as some of the more critical researchers in the health-care-reform debate point out, particular kinds of evidence are "privileged" over others, namely statistics in designated abstract categories, plus clinical data (the "gold standard" of which is the randomized control trial, which is very expensive and therefore doable only by those institutions with lots of money). As well, while statistical measures of efficiency carry the aura of scientific objectivity, the more skeptical researchers also know about "data gaming," or "data massaging." Hospital readmission rates too high? Well, adjust the parameters so that anything beyond forty-eight hours after a patient is discharged is counted as a new admission. Emergency-room waiting times too long? Change the start time to begin only after the person is giving their history to the admissions clerk, or whatever. (In some research on hospital cleaning that has been outsourced for cost-cutting and efficiency, the part-time and contract workers refer to their work-sheets as "lie sheets" because so often they just tick and sign the work no matter how or whether it is done.)

The "evidence" that's needed for accountability in institutions like hospitals to honour the spirit of their core values and ensure they're reflected in practice clearly isn't just in the data. It's equally in the stories that patients, doctors and nurses have to tell, including not only their voice making a difference, but their empathetic ear.

"You can't do nursing properly if you don't listen," one of the nurses tells me in Toronto.

"With your heart as well as your mind," another adds.

"And that takes time," the first nurse says. "It takes time to observe and listen."

Doctors have traditionally responded to nurses' intuition, even when there's no change in the quantifiable health indicators such as the organs' vital signs.

The public trusts it too. "The patients feel it, or read it," one nurse added. Perhaps it's the rhythm of attuned attention that knits all those fragmented tasks and technologies into the comprehensive whole of a hopeful, healing relationship.

Somehow the officials charged with restructuring health care and ensuring its effective "delivery" across the country must find a way to trust it as well, perhaps by exploring the use of narrative and dialogue as accountability methods to complement the data. Perhaps by letting more health-care discretion and decision-making return to the time and space closest to where the reality of empathetic engagement can be felt first-hand. The just-in-time delivery of resources will be both efficient and effective in a context where the people involved can make sense of what's going on in the here and now, combining timely moments of fast diagnostics and surgical intervention with ongoing teamwork and patient-care relationships in which they can judge what action is most appropriate under the circumstances. And that takes time, including the slowed-down pace of one human being focussed on another, caring for each other in collaborative relations and emergency-room teams while also caring for the sick.

The medium is the message. An ethic of caring doesn't spring fully formed from some branding consultant. It's embedded in relationships, with walking the walk in the practice of health care day-to-day, with others holding themselves equally responsible for the health of society and all its members. There needs to be a balance in the methods and language of accountability so that the ethos of empathetic caring can be meaningfully delivered. Otherwise, the "voice" behind the curtain could soon be a beep: a statistical indicator of vital signs, and nothing more human than that.

Six

MINDING THE
COMMON WELFARE

"Society is not paying enough attention."
WITNESS at coroner's inquest into Jordan Heikamp's death

*"A fundamental of social work is that you change
a client's behaviour through a relationship. You can't
order someone to do something."*
SOCIAL WORKER

ORDAN HEIKAMP was born in Toronto weighing four pounds, six ounces. Five weeks later he died, weighing four pounds, two ounces. "Little more than skin and bones," the coroner's inquest into his death was told.

For the duration of his brief life, baby Jordan was officially in the care of the social-welfare state, first in a hospital and then in a shelter with a social worker assigned to his mother. The jury heard much exonerating evidence: observations jotted down in logbooks, phone calls made and duly noted in case files. But no one interpreted the fragmented comments, put together the scattered observations and took hold of the situation. The coroner's lawyer was stunned at what he called "the pas-

140

sivity" of all the caregiving professionals putatively involved in the case. It was as though they were present and not present at the same time: connected to the case, yet disconnected.

The coroner's jury of three men and two women ruled that baby Jordan's death was a homicide. They determined that the cause of death was "chronic starvation" brought on by "neglect." But it wasn't neglect actively and culpably associated with any one, two or more individuals. Instead it was dubbed a "systems failure," at least by the head of the Catholic Children's Aid Society. Equally, I think, it was failure on the part of a society that would leave so much responsibility for people's welfare to a fragmented, indicator-driven social-welfare system that a baby could literally waste away and die before that system's very eyes and in its official hands.

I found myself asking a lot of questions after baby Jordan's death. Is the welfare of society being disabled? Or are we losing our ability to look after it? Are we as a society no longer paying attention to what matters on the ground, to what matters to us as a social commons? Are we no longer able to pay attention to the point that the realities of data can take over the space and time in which dialogue and an empathetic eye and ear used to make a difference? Are we becoming a truly dysfunctional society, unable to look after our own? The questions reared up because babies are society in a microcosm, small and fragile enough to hold in our hands. In this case, in the palm of one hand.

Babies aren't just the link in the chain of life from one generation to the next. They're the living, breathing point where one generation passes on life to the next, palpably demonstrating that we are more than the sum of our individual selves. They're also proof that the chain that is society is an organism. As philosopher Mark Kingwell argues, we are always oriented to the world and each other, even in our rationality. It's the mutually oriented web of relationships that sustains us as a society. As these weaken, society itself can starve.

THE TRANSFORMATION OF SOCIAL WELFARE

Historically, social-welfare work has been rooted in an ethos of human compassion and justice, with justice, like compassion, understood as something qualitative and context-specific. When social welfare was left to local communities—to churches and other charities—it depended on the character and justness of that place and of those people. And those qualities were often sorely lacking. At "pauper auctions" (the last recorded Canadian example of these having been held in Sussex Parish, New Brunswick, in the 1880s), local farmers and others in need of cheap labour would bid against each other for the destitute men, women and children being kept by the county. They wouldn't bid the price up, but rather down. Wherever the bidding stopped determined how little money the person's self-deputized keeper would ask from the county to help support this poor individual. The idea was that the paupers were to fill in any gaps through labour. And if they sickened and died, well, the person looking after them got to pocket any savings left over.

Increasingly, charitable organizations took charge of the public well-being. By the mid-twentieth century the social-welfare state had placed public welfare on a plane of universal rights and entitlements, though some critics, like philosopher George Grant, sensed a chill setting in as rights as abstractions replaced moral rights and duties embedded in relationships and communities. Social welfare was redefined around bureaucratic rules for allocating various entitlements, such as extra money for food if a destitute woman was pregnant and unmarried; still, the new standards continued to be interpreted by engaged professionals in the context of face-to-face dialogue and counselling sessions with people on welfare. Social workers had enough discretionary time and space to "reconstitute the social," in other words, to be present with people in the mildewed apartments where they lived, the bleached-out laundromats where they washed their clothes, even in the dark alleys where some of them "dived" into dumpsters for things to eat, to sell or to trade. The system was open enough even to allow for

some expansion during the 1970s and '80s into shelters for battered women and homes for teenage runaways. Since then, in the
name of public-spending cutbacks and accountability plus a specious "war" on welfare fraud, the system has been radically restructured and, in the process, closed in on itself.

Jordan Heikamp died during the final total-systems phase of restructuring social welfare in Ontario. He died six months after a
government contract was issued to Andersen Consulting to develop a comprehensive system for social-assistance assessment
and delivery, plus tracking and reporting for accountability audits.
This was a joint venture between the consulting firm and the Ontario Ministry of Social Services, and it represented one of the
largest public-private partnership agreements initiated under the
government's new Common Purpose Procurement program. One
of the program's principles stated that the "private-sector vendor"
selected to work with public servants shall "share the investment
in and risks and rewards of the project." The idea was that in return for developing what the partnership agreement called a "cost-
effective" delivery system, the company would receive a share of
the cost savings achieved through its tighter and more technologi-
cally controlled eligibility assessments for people applying for, and
often in desperate need of, assistance. Henceforth, these people
would be called "clients."

Andersen Consulting (since renamed Accenture to distance itself from former parent company Arthur Andersen, which was
heavily implicated in the Enron accounting scandal) has carved
out a niche for itself in reorganizing social-welfare services into
more businesslike systems. It centralizes the management and
definition of work into one information-management system.
Through it, the people seeking help from social services are standardized into computer-coded "needs" and "outcomes," with set
rules for working out the connection and quantifiable performance measures for tracking the results. A second and related
theme is to put as much cost-related decision-making as possible

into the computer programs rather than leaving it to humans, whose hearts might be moved to find a way around the rules. As one example, the "eligibility engine" software for calculating whether people should receive social assistance, and at what level for different categories of life's needs, is entirely based on predetermined coded inputs. Cutting off someone's welfare payments is similarly based on standardized coded inputs and happens automatically, though it can be appealed. Having software control most of the decision-making around who is given social assistance also provides a digital trail of accountability. By the by, it can also help in realizing the cost savings so central to the success of private-sector involvement. If people don't meet the computerized eligibility requirements, that's so many more dollars saved.

(The Children's Aid Society [CAS], a social-welfare agency specifically devoted to the physical and emotional well-being of children, was not included in this re-engineering project; however, the upheavals the restructuring caused and the rigorous reporting priorities it introduced reverberated through every area of social services where cutbacks and staggering caseloads had already raised stress levels considerably.)

Having developed standardized categories both for need and assistance provided, the consultants refined these into a set of prescribed questions for assessing need and eligibility, interviewing a host of social-service workers and poring over their files to get a picture of the complexities involved. Compressing the differences and subtleties of human experience into standardized terms of reference, they created abstract indicators of need, risk and financial self-sufficiency, which could be applied equally and objectively to all. The related questions were intended to save a lot of time as well; workers need no longer listen to each person's story, pulling out salient bits of information and interpreting them in light of each person's situation. The result is that under the new system social workers—who are now called "income-maintenance officers" or "eligibility-review workers" or simply "case workers"—no longer

144

interpret or judge things for themselves at all, except under exceptional circumstances of extreme risk.

By 2001–2, the key components of the integrated service-delivery system were being implemented. These included "smart cards" issued to social-assistance recipients, with a 1-800 telephone number for accessing the interactive voice response (IVR) system any time, twenty-four hours a day, seven days a week. The system works as follows: Having dialled in using a push-button phone, cardholders enter their nine-digit identification number followed by the four digits of their own personal identification numbers, or PIN. Once inside the service-delivery system, they're invited to press "1" to know the status of their file (either "ongoing," meaning it's active and the cheque is in the mail; "suspended," the cheque is on hold while Social Services investigates possible overpayment or fraud; or "terminated," it's closed and there will be no more cheques). If they press "0" a real case worker in a call centre comes on the line. Among other things the worker can explain how to initiate the process of appeal if benefits were denied or if the file was suddenly terminated.

People needing help and applying to receive social assistance for the first time use a modified version of the same IVR system. It's called an intake screening unit (ISU), and anyone applying for social assistance in Ontario is expected to use it. The access point once again is the touch-tone phone: Once callers enter the system, pressing "0" routes them to a person in a client-contact centre who takes them through the list of standard intake and eligibility questions and tells them what documents to take to the nearest People Services office for verification.

I found myself wondering about this system in relation to baby Jordan's mother, Renée. Raised in small-town Campbellford, Ontario, she'd run away from home at age sixteen to join a circus. She ended up on the streets of Toronto, and pregnant. But when she phoned home her mother was in the throes of moving to North Bay, Ontario, with her new husband and, for whatever reason,

Renée stayed put. I don't know whether she phoned an ISU at that point. Still, to get some idea of what it might have been like, I dialed in myself. After a few more calls, I arranged to meet with one of the workers. To protect his job, we agreed not to use his real name. I'll call him "Sandy."

A FACE BEHIND A CONTACT CENTRE

The call-centre ISU where Sandy works consists of sixty workers spread across three locations, though all are served through one call-distributing system. So, he says, where he's physically located "isn't significant." It's as insignificant as the particulars of where I called from, I think, and what I look like: dirty or clean, hungry or well fed, hung over or healthy. From whatever the source, the call goes into Sandy's headset at his computer cubicle workstation via the centralized computer-dispatch system. With every call he follows a script of between forty and fifty questions, to which he duly inserts the answers into standard fields displayed on the screen in front of him. A lot of the questions have to do with finances and assets.

As we go through the questionnaire, I ask him if he's affected by the more businesslike language in the system now, all the codes and computer programs, like "eligibility engine."

"It sounds pretty crude, doesn't it? An eligibility engine that crunches the data to determine whether this person is worthy of assistance or not."

"What do you think?" I ask.

"What do I think? Umm, it really hasn't done very much for efficiency, and people are beginning to realize this."

"How so?"

He explains that people still have to visit a People Services office with documents to verify everything and sign various forms and statements. "So the whole process is taking twice as long as what it used to, because of the verification process."

146

Meanwhile, he's got that queue of calls waiting and he's trying to meet a best-practices time for processing each call. He explains, "They're still working on the durations here, in terms of how long it actually takes from beginning to end. That's actually happening right now, in our office. We're piloting this. They're number-crunching—and they have us coding in the type of call, too."

"It can become very impersonal," he continues, "when you're pushed to deal with a number of individuals that are showing up in your queue. There's pressure to go through it quickly. And the faster you go, the more impersonal it becomes. When you're just dealing with numbers, it's very easy to forget that you're dealing with people. You're dealing with lives; you're dealing with feelings, with goals, desires. Many of the people are arriving at the welfare doorstep with their dreams and desires crushed because of their financial situation or their circumstances, abuse or whatever their reason is for not being able to hold down a job.... It becomes a matter of *processing* the individual rather than really looking at them as a human being. And it's the time constraints. The drive to do things faster and more efficiently. More people in less time."

Sandy is one of the younger generation coming to social work through community-college programs and training in information technology (IT). In his case, the IT diploma is framed beside a bachelor's degree in religious studies.

I ask whether he's ever affected by the squeeze when he's been talking with someone who's having trouble getting by. He tells me he has the discretionary leeway to push the "not ready" button to block further calls until he's ready for the next. "I think you need to use it sometimes, to maintain your own sanity," he says. "It's a little too much at times. The repetitiveness of it, it's mentally exhausting. The same questions; often the same responses."

I wonder if the same goes for people at the calling-in end, people who can barely keep from crying and instead find themselves asked for their social insurance number, their bank account number. When they are referred to a local social services office, I wonder what they feel when there's a computer on the desk between them and the worker they've come to see. A survey of social-services workers in Ontario found that 67 per cent reported having been subjected to verbal abuse by clients, 46 per cent had been threatened and 24 per cent had been physically assaulted. Some social-assistance offices have installed thick Plexiglas windows between worker and client; I can only imagine that they exacerbate the problem while nominally containing it.

I ask whether trust has a chance to develop in this disembodied standardizing milieu. "Trust, that's an interesting concept," he says, then talks of how these conversations often begin with enormous suspicion on the part of the applicant.

"I think we're viewed as income-maintenance officers. Well," he pauses, smiling across at me, "that is the term that the province uses for their social-services case workers. It's a little colder. It's maybe a little closer to the reality of what we do.

"We ask them a lot of questions that are rather personal, especially single mums. We ask about the father of the child, and his whereabouts. And one of the strangest questions is, 'Are you currently living common law with a spouse or a same-sex partner.' It's very... uncomfortable for a lot of people. It's very intrusive. Is this a roommate or a lover? If it's a lover, then we have to take the application in a different way. Your monthly entitlement is affected if you're a couple. It's slightly less."

Still, he continues, "I think the leeway is still there to deal with individuals who are in dire straits and do require a little bit more personal attention than your standard eligibility assessment."

I feel better hearing this, though it assumes that the case worker notices the need and is moved to offer that extra attention. I ask Sandy if there's a certain pace that empathy requires.

"That's a good point. Yes, it does take time to get to know another individual."

"And to give them a chance to sense that you care?"

"Exactly. And it certainly doesn't appear in a list of forty questions. 'How are you?' is not one of the forty or fifty questions we ask."

"So this empathy can just sort of disappear?"

"Yeah. The pace has to be slowed down; that's true for any relationship."

I wonder aloud whether this is how people can "disappear."

"Sure. It has to be. People fall through the cracks."

But then he tells me it's dehumanizing for the case worker, too. "It becomes information overload after a while. It can be desensitizing.

"When you become an income-maintenance officer or an eligibility-review worker, you become less than human. And unless you fight it, it will burn you out."

We sit in silence, a bit stunned by the import of that statement and the vehemence with which Sandy has just made it.

He continues, "Me, I make a conscious effort to keep it human; I think that saves me. I try to slow things down a bit between questions. Find out a bit about what's going on in that person's life, a little bit of the social-work sort of thing. My own compassion, if you want. But it really has to be inserted. I can't think of a better word than that. It has to be added back."

I ask this still hopeful young man whether the system gets to him, despite all his mitigating efforts.

"It certainly does drain you. If it robs you—well, it can only rob you if you let it rob you. But it certainly does drain you. All the pressure there is to produce in terms of how many applications you process. All those words that come through in a day. Yeah, it comes down to not having enough time with these individuals to establish a rapport, to find out what their needs are. You can't do that much for them, except make referrals. But still you're constrained in how much you can do.

"People are reaching the saturation point. The time to reflect is not there, and people feel that—the time constraints. And they (management) want you to reduce every situation to a three-digit code. It's not very easy to do. 'What category do I put you into?'

"And your calls are monitored. And the time that elapses between calls. We classify a call according to codes now: 'General inquiries.' 'Active client' queries; for example, who their case worker is. And there's one... this is an interesting terminology we use—we get a call from a sixteen-year-old. Sixteen-year-olds we handle with kid gloves basically because of their age, and the delicate situation they might be in: they might have been kicked out of home and they're calling us from a shelter.

"We do not use the technology in that case. What we do, it's called a 'warm transfer,' which means that we put them through to a human being on the other end of the line for an appointment."

THE VALUE OF HUMAN CONTACT

Renée probably entered the system as a warm transfer. Seasoned social worker Denyse Roberts would have had no trouble identifying with her, either, if the pregnant young woman had been transferred to her care.

I met Denyse through lovely serendipity. She's a friend of Gail Martin, who transcribes my interview tapes and who, after doing Sandy's, insisted that I talk to this woman. Denyse has seen it all, been through all the changes, Gail said. I discovered just how true this is after I'd spent several hours at Denyse's home.

Denyse tells me: "I lived on the street for almost three years with all of what that entails. It was horrible. I was put in jail in Vancouver when I was fourteen because I was using someone else's identification and I was charged with a crime and convicted and sentenced to jail, and it came to light that there were two of us so I ended up being shipped back home. I was connected with a juvenile social worker at Bronson and Somerset [in Ottawa]. Miss Hardy. I don't know what happened to her.

"She'd just show up and sit," Denyse says, referring to the period before she was ready to receive the professional helper's presence.

"I remember her saying to me, 'There's something special about you. You know, you can choose to do this.'"

Slowly, Denyse opened up to Miss Hardy. She got back in touch with her mother, in time to be there during an illness that over the next three years would take her mother's life. Denyse attended adult high school, then went to college and university to become a social worker. After graduation she started working at the Youth Services Bureau, where she had the "incredible luxury," she tells me, of having a caseload of less than ten.

"I was on call twenty-four hours, seven days," she says, explaining that she'd get all kinds of calls. "'I'm in a snowbank and I can't find my boots,' that kind of thing. I met some incredible kids."

I ask if she felt she was ever able to make a difference with these kids.

"I think so. The time I was able to spend with them built up a level of trust that they would never have gotten otherwise. When you are sitting in a car at 4 a.m. with someone whose feet are frozen and they don't want to go back to where they are staying, you can go a lot of places from there."

"You went out into that snowy night?"

"Yes. Or being in an arcade playing games with these kids. They talk about stuff that they don't talk about when you are sitting behind a computer. These things would never, never come out. I've had kids really save me from other kids. Kids who stepped in front of me and pushed someone aside because something was being thrown at me. A moment when one of my clients called and asked me to take her child out of her house because she was afraid she was going to hurt her. That was really powerful. I don't think she would have called me if we didn't have the relationship we had."

We talk about timing, and taking the time to read another person and tune in to the rhythm of how they talk and go about things.

"You can't just be present and expect communication and trust just like that," she says.

151

Is there time for that now? I wonder. I ask, "How big is your caseload now?"

"One hundred and twenty," she replies.

"So the time to be present, to develop a human empathy… " I trail off.

"… is practically non-existent," she says. "It has been squeezed out of the system."

HERE IS THE ROOT of the crisis in social welfare, I think. There's neither the time, nor the congeniality of shared space, for someone to tell their story. Nor, at the other end of the line or on the other side of a Plexiglas-partitioned cubicle, is there much chance for an empathetic ear to hear not only the nature of the problem but also the possibilities. There is little or no scope for building trust and rapport and a genuine relationship out of which the person can start to hope, to move forward. It's not so much that there's absolutely no time left for this, social workers are yet another set of professionals squeezed between the realities of life and data on the screen, trying to serve both in the tight time frames provided. As Denyse and others tell me, the committed ones insist on making time for their clients, even if it means staying late to get their reports done. Angela Martin, the social worker initially charged with criminal negligence in the death of baby Jordan, was typically working until eleven o'clock almost every night.

Yet the standardized codes and data fields are so pervasive and enclosing that even with the best of intentions it's hard to revive something authentically human, especially when you're so stressed out and exhausted you can do little more than go through the motions. Meanwhile, the reality on the ground is that clients are being disappeared and left unattended—almost abandoned. An Ontario study of child-abuse cases between 1993 and 1998 found a 58-per-cent increase in abuse cases among families that had had previous contact with a child-welfare agency. A lot of peo-

152

ple weren't paying attention or, if they were, didn't sustain it for long enough to make a difference.

BABY JORDAN AND THE DIGITALLY DISAPPEARED

So, what happened in the case of baby Jordan? Renée drifted on and off the street, in and out of a teen shelter, through much of her pregnancy. Then the contractions started and she was admitted to hospital in a rush. The baby was coming well before what was thought to be its due date. Jordan was born on May 18, 1997, by Caesarean section, and was immediately placed in an incubator. Three days later—and eight days before she was due to be discharged—Renée checked herself out of the hospital without a doctor's consent, though she left Jordan to receive continued care. One of the nurses worried that Renée hadn't yet learned to feed her baby. Still, as she told the coroner's inquest into baby Jordan's death, she reasoned that Renée would get the necessary help at the women's shelter where she'd arranged to live. The nurse thought there were nurses on staff; in fact, there were counsellors on staff, some with nursing backgrounds.

On May 29 baby Jordan was released from hospital, into Renée's hands but under the joint supervision of Angela Martin, a social worker with the Catholic Children's Aid Society of Toronto, and a local women's shelter. At the time, Martin had a caseload of thirty-eight and had been instructed by her supervisor "to ensure the safety of clients and not to do more than that." The doctor had made an appointment for Renée to go and see him with the baby, an appointment she apparently never kept.

Martin touched base with her new client, asking if she was going to see her doctor and, encouraged by the young woman's assurances that she would, made plans to do a formal "risk assessment" on the baby (to determine whether he could safely be left where he was) as soon as she could work it into her already overloaded schedule. Meanwhile, Renée took care of her baby as best

153

she could, putting a little liquid formula into a baby bottle and liberally topping it up with water.

At the coroner's inquest, a doctor testified that the bottle was probably filled two-thirds with water. At the time of the baby's death, there was absolutely no food in his body, and his organs showed signs of "extreme wasting." The paediatric pathologist who performed the autopsy told the coroner's jury that Jordan would have shown signs of malnutrition at least ten to fourteen days before he died. Sunken eyes and wrinkled skin would have been two of the telltale signs.

But nobody, it seems, noticed. As Renée's lawyer told the inquest, "Renée's irresponsibility, unreliability and reluctance to learn should have been obvious to anyone who paid attention.... Her incompetence [should have been] plain to anyone who asked a single hard question. No one ever did." Baby Jordan died before Martin could do the post-hospital risk assessment and, even if she'd determined that he was "high risk," there was a five-week waiting list for the high-risk assessment committee to consider a case.

Shortly after this case, the Ontario CAS conducted an internal time audit and discovered that its social workers spend up to 85 per cent of their time on computer work and other case-administration matters, and only 15 per cent of their time working directly with children and families in crisis. (Nationally the situation isn't much better, with 30 per cent considered the average amount of time a social worker spends in the field, yet with an increasing number of children at risk coming into their care.) A survey of CAS workers in Toronto, which confirmed the Ontario-wide findings, concluded, "We are concerned that the system is becoming dehumanized as workers take shortcuts in communication [with each other]. Workers are tending to refer to family situations by numbers and letters rather than by name." And a Dutch study reported, "The use of a computer during contacts with clients seems to encourage... workers to dissociate themselves more easily from an individual client who acts in a way that is unsuited to the requirements or expectations of the bureaucracy. Case workers using a computer during contacts indicate that in case of disagreement with clients, they are less inclined to go into the problem, talk it over and search for a solution."

Bill is a Toronto children's aid worker who made it his business to be in court on the day in April 2001 when the coroner's jury returned its verdict about baby Jordan, dismissing all charges and implications of criminal wrongdoing against his colleague, Angela Martin. He was stunned that the only fault they found was that she could perhaps have phoned the doctor's office to check that Renée had taken the baby to see him. Bill went home resolved to quit social work despite twelve years of service in a field that he clearly loves. I caught up with him when he was still considering the drastic move. He felt he should quit, he told me, because it could have been him. He could have been the one who missed making that phone call, a call he would have had no way of knowing was important because he was so stretched with overwork that it would have been just another call on the list. He realized that day that all the bits and pieces of pseudo-connection with his clients left him feeling essentially disconnected from the people he thought he was there to help, and they effectively disappeared.

At the end of the day, it's not just being stressed and overworked that bothers Bill. Nor is it the priority given to completing reports on time, even though this often means using Code 9 for "insufficient information" or filing them on time but with the "not complete" box ticked off. What concerns him most is losing a feel for the people behind all those forms, files and phone calls to be made or returned and logged for the record. It's being dissociated, disconnected from his own humanity, his capacity to be engaged, to be present, to empathize and relate. The deficit of attention deepens into a deficit of feeling, which can translate into neglecting the people he entered social work to help.

He explains, "You can spread yourself so thin over the caseload that you can't serve or get to know anyone properly. You are out of touch with them really.

"My anxiety level is so high that there isn't much room for empathy. You are so focussed on yourself and your survival that it is hard to be sympathetic or empathetic. You have to remind yourself to do that, to take a moment for this person and listen to them."

It's a form of sensory deprivation, I think. They as people, their bodies, the body of their experience and the realities of their lives, all of that's out of the picture, I say.

"That's right," agrees Bill's colleague, Rod, who works for CAS and has just joined us over lunch. He laughs. "We have an income reporting statement that still comes in once a month and clients have to report their change in circumstances. Now it has become sort of a cynical black humour that we try to identify the stains on the income statements—you know, is this baby poop or peanut butter?—because you try to derive the lifestyle of the client from the coffee rings on this statement that they must be filling out once a month at their kitchen table. You could have gone and actually observed the situation they live in.

"We still have a requirement to go out and do fieldwork, but we go out and do it in a hurry, because you don't have enough time. You're almost faking it. You're thinking, 'I need to ask these eleven questions because that is the form I have to fill out when I go back.' So you make sure you cover them, and once you cover them you just want to get out because you have other things to do. You've got what you need, so now you can go back and fill out the form and get on with the other dozen pressing things you have to do."

"So the person might be telling you something that in fact is important, but you are really not there," I suggest.

156   "Yes," Rod says. "You are looking for the answer to one of those things on your list and you are so worried about everything else you have to do that you are not listening to the peripheral stuff or the stuff that goes into painting a real picture of who they are as

human beings and what they are going through as human beings. You fill out forms because you have had some technical questions answered or you are writing 'insufficient information' and sticking in your comments, but you don't feel you really know what is going on. You don't really know if there is a problem and you can be nailed like that other social worker [Angela Martin] was nailed."

Hence the anxiety, Bill adds. Case workers aren't just afraid of ending up with a death on their caseload, as they go through the motions, covering the blanks with Code 9s et cetera, they're afraid of unwittingly allowing such deaths to become almost inevitable as they become detached, numb and out of touch with the human realities involved.

"It's a moral crisis," he says. "You feel you are carrying an ethical deficit." A deficit of compassionate engagement.

BILL'S COMMENT stays with me all the way home on the train. An attention deficit becomes an ethical deficit as the new system not only fragments social-welfare workers' attention but abstracts it from the context of real life. They can end up abandoning not only the people who need help but themselves, as they withdraw (and are withdrawn) from daily relationships that engage them in minding the welfare of society effectively. It's both an institutional forgetfulness and a self-forgetfulness. And yet the more rushed and stressed they are, the more numb they can be to this fact and its consequences. Helpless, needy babies can become simply more data, not through wilful but an almost engineered indifference, yet so subtle and embedded in daily routines that it's almost irresistible. A set-up for a dysfunctional society? We could sleepwalk right into it.

$Part$ THREE

SOCIETY:

Bringing a Crisis of Accountability
and Meaning Home to Roost

Seven

CHILDREN'S TIME AND
ATTENTION DEFICIT DISORDER

*"Temporal organization may be the fundamental principle
underlying the earliest social integrations."*
BEATRICE BEEBE, Of Speech and Time

*"Wisdom has its own curriculum: conversation, thought,
imagination, empathy, reflection. Youth who lack these 'basics,'
who cannot ponder what they have learned, are poorly equipped
to become managers of the human enterprise in any era."*
DR. JANE M. HEALY, Endangered Minds

"APRIL 30: Donald is sitting propped up in the seat
beside me.... He's watching his hands as he or-
chestrates them through the air. Now he lets out
one of his melodic screeches. He looks my way. His eyes
are searching for mine, a smile ready as I turn to look. I
smile. He immediately grins. He makes another little
'ayey' sound. I smile again. He grins again, this time
making another ayey noise as he grins.

"If he's doing something and hears me chuckle, he
looks up immediately, ready to join in my laugh as soon
as our eyes meet."

More than twenty years after I wrote these words, in
a journal I started after my son, Donald, was born, I still

remember the thrill of those first few months with him. Our bodies were so in tune, our attention so focussed on each other, that a touch or a look was enough to convey whole volumes of love and connection. I've dug out the journal now with the thought that shifting to virtual realities at work or at play might not matter as long as we can still "get real" and engage with each other face to face. What's important is how we relate to one another and are present in society. Research shows that this begins at the earliest stages of life.

INFANTS, YOUNG CHILDREN AND THE ROOTS OF SOCIAL INTEGRATION

In a speech I gave some years ago, I suggested that the umbilical cord might be considered the first line of communication. I had in mind how a woman's stress is biochemically communicated to the child she's carrying, but mostly I wanted to restate the obvious, that the body is the first medium of cultural expression, as the French word for language, *langue,* or "tongue," so deftly reminds us. The body is the first medium through which we articulate ourselves and engage with others. It is the point where we dwell in the world, in society. Equally it's the site where the world dwells in us, through our senses taking it in. It's through our bodies that we come alive, taking up conscious residence in the world as we open our mouths and utter (or "outer," as Marshall McLuhan once rephrased it) our thoughts and feelings. Communication is as much non-verbal "body language" as it is verbal conversation. It includes the pauses, the attentive listening, the give and take of dialogue in relationships that, as they flourish, in turn nourish society in the best humanist tradition of the word.

When I talked about umbilical cords that day, I hadn't yet learned about fetal clocks and how mothers transmit their circadian rhythms to the unborn through that same cord. The key, it seems, is a cluster of what scientists are now calling "pacemaker" cells in the brain's hypothalamus. Shortly after these cells

162

develop, a circadian rhythm starts up inside the fetus, entrained to the mother's.

Basic rhythms of communication in infants build on this by setting up "rhythms of knowing" and synchronizing with others throughout a person's life. Babies learn about timing, about the basic pattern of action-reaction, from their mother's (and sometimes father's and other caregiver's) repetitive actions. Researchers describe it as "kinesic rhythm." By about three or four months after the birth, mother and child have created a "split-second world" in which they're so sensitive to each other's non-verbal communication that each is in effect "responding" to the other within less than half a second. This response is called "co-action," which implies that babies aren't just reacting but are predicting and anticipating when to make the next move. According to researchers, as babies start to take some initiative in this dialogue they are shifting from solitary, egocentric beings to participants in a larger social context.

Some researchers in the fields of linguistics and early-childhood development speculate that human language evolved from this pre-verbal mother-child interaction. They also think that the capacity to communicate as adults grows out of this rhythmic attunement to each other, which linguistic scholar Noam Chomsky calls the "universal grammar" of learning one's mother tongue. In short, researchers have new respect for "baby talk." Characteristically high-pitched, repetitive and slowed down, with exaggerated emphasis on rhythms, it is now called "motherese." Equally important, says McMaster University psychology professor Laurel Trainor, is the "emotional content" of this speech, which encourages infants to pay attention.

Although infants engage mostly in solitary free play, which helps them develop a kinesthetic sense of the world through direct and sensuous involvement with it (and apparently also sets the stage for learning math that involves abstract manipulations of space), toddlers begin to engage in social play.

When children are young, active, hands-on experiences such as playing, exploring and talking are important, as are spontaneous improvisation in made-up games and dramas and listening to someone tell or read a story. They all extend the rhythmic reciprocity of dialogue and the engaged attention of listening, which, according to educational psychologist Jane Healy, helps lay the foundation for literacy. She stresses listening because, with sound, the sense and any information content tends to unfold slowly with the articulation of notes, syllables and pregnant pauses. Listening cultivates sustained attention by requiring it, whereas sight allows us to grasp a visual image in an instant. Citing others' research, Healy argues that the ability to read builds importantly on a child's "phonological awareness," that is, "the ability to identify, remember and sequence the sounds in words."

I am struck by how much the rhythm of reading a children's story aloud echoes the earliest rhythms of attention between parents and their infants, and how much these seem to reiterate the primal meaning of time, and pre-modern conceptions of it: the pulsing of the heart, the ebb and flow of the tides, the wax and wane of the moon and the seasons. I'm also impressed by how this capacity for sustained, mutually attuned attention in time is linked to an emergent sense of self. Canadian psychologist Dr. Virginia Douglas concluded a 1972 paper that helped lay the groundwork for understanding attention deficit disorder (ADD) by quoting the renowned American psychologist William James: "If a brief definition of ideal or moral action were required, none better would fit the appearance than this: Attention with effort is all that any case of volition implies. The essential achievement of will is to *attend* to a difficult object and hold it fast before the mind."

In tracing the roots of attuned interpersonal rhythm, psychologists also stress the influence of the conditioning environment of the home and the social context of love. In *Childhood and Society*, for instance, psychologist Erik Erickson emphasizes the importance of "the mother's focussed attention and care." (If he were writing today, I hope he would credit fathers and hold them equally responsible, not just for their own sakes but because research shows that kids whose fathers play with them in a sensitive, supportive and challenging way at age two tend to form closer, more trusting relationships with others at ages ten and sixteen.) Canadian psychotherapist Dr. Gabor Maté says much the same thing: "The direct, calm interest of the caregiver first kindles the infant's own interest in the world and helps to organize his attention." And later: "The skill of attention that begins during the initial stages of brain growth [is based on] the secure attachment/attunement relationship with the primary caregiver.... Attention and emotional security remain intertwined throughout childhood."

I flip back through the journal I started after my son was born on January 10, 1982, and am amazed at how much my words echo those of the experts I've just, for the first time, been reading.

"February 2: Today I was talking to him so naturally, actually enjoying him even in his fussy moments. Then I danced around the house with him held slightly aloft in my arms. He's so attentive. He listens and seems to respond. He's like a violin string quivering he's so responsive to my vocal cords, and whether they're tense or relaxed.... "

"March 15: There is a sequence that begins now with me sticking my tongue out at Donald. He looks at me, then he sticks his tongue out. I grin. He grins, his eyes twinkling. Tears spring to my eyes, and I laugh with delight. His grin spreads to the point that I can see part of his gums, and he lets out a gurgle bordering on a laugh.

"He's been smiling for a good three weeks now, and so attentive!"

The word "attention" comes from two Latin words, *ad* and *tendere*. Together, they mean "to reach toward." Attention's root, then, isn't just an active verb. It's an interactive verb, of reciprocal reaching out and response, both nested in feeling and the rhythms of shared time in a relationship. Attention, like attunement, is the *sine qua non* of bonding, not only between people but with the larger living world that sustains us all. Equally, it seems critical to feeling rooted and secure in life. (Research in the relatively new field of epigenetics, or what turns genes on or off, has found that attentive mother rats tend to lick and groom their babies a lot, which seems to activate the genes that regulate stress hormones, because when they're mature these rats tend to stay calm under stress when their unlicked lab mates don't.)

It's almost as though early-childhood development—all those clumsy, imperfect gestures of love between parents and their newborn babies, from stroking and kissing to song, talk and games—is a second stage of delivery, channelling people into effective participation in the world. These mutual gestures jell into a habit of attentive social engagement at home, at daycare and later in the classroom and playground at school. Yet we live in an increasingly attention-deficit culture where attention itself is treated as a resource to be marketed, and various devices vying for that attention—televisions, computer games and the like—are everywhere and turned on for large chunks of people's waking lives. Attention can become so indistinguishable from just another deliverable that the mutual engagement implied by its root meaning is distorted or even lost. The danger increases if the family home is no longer a slow-zone retreat from the madding crowd, if parents' own attention is scattered across a thousand demands and errands and they just don't have the time to be really present with, and focussed on, their kids.

ATTENTION DEFICIT DISORDER AND
THE THREAT OF SOCIAL DISINTEGRATION

Attention deficit and hyperactivity disorder (ADHD) and attention deficit disorder (ADD) hit the news in the late 1990s. Recognizing that too many kids are unable to concentrate, pay attention, sit still and not disrupt the class with impulsive behaviour—the classic characteristics of both—education and health officials began to take note. Although the U.S. accounts for 90 per cent of the prescriptions for Ritalin, the medication most commonly used to treat ADD, Canadian physicians wrote nearly one million prescriptions for an estimated 20,000 patients in 2000. By 2001 the increase for the decade had been pegged at 600 per cent.

Scientists are divided over the cause of this disorder, and many wonder whether medications such as Ritalin are effective or whether the drug itself was used to coin the illness. Sure enough, genetics is fingered as the culprit in ADD, though opinion seems to side with the messier, more complex picture associated with childhood and the social context of our life and times, rather than with a single deterministic root cause.

The term "attention deficit disorder" formally entered medical dictionaries in 1980, but it really dates from 1972 when Virginia Douglas took a hard look at the literature that had accumulated on hyperactivity in kids and combined it with her own (and colleagues') research to write a paper shifting the focus to attention. In twenty closely argued and well-documented pages, she set out her premise: It's "[b]ecause of his short attention span [that] he tends to flit.... As a result, his behaviour is often fragmented and disorganized." Without actually using the words "attention deficit disorder," she concluded that the core problem was kids' inability to "stop, look and listen." "These youngsters," she noted, "are apparently unable to keep their own impulses under control in order to cope with situations in which care, concentrated attention or organized planning are required." As a result, not only can they not sit still and learn, these kids often have no steady friends, and a sizable minority have a history of anti-social behaviour and even serious acting out. Thirty years later Douglas, a professor emeritus at McGill University in Montreal, is still doing work in this area, though she says that if she was asked to coin the term now she wouldn't call it ADD. She'd call it "a self-regulatory disorder."

Gabor Maté is not only a respected authority on ADD, calling it "a problem of development," but is one of its statistics. He was born of Jewish parents exactly two months before the Nazis took control of Budapest, where the Matés lived in 1944. Of course he doesn't know his mother's state of mind while she was nursing him through his first months, but one can imagine her being distracted, to say the least.

"The human being is not genetically programmed," he tells me on the phone from Vancouver. "The potentials are genetically set, but in any particular circuits in the brain, whether it has to do with hearing or vision or attention regulation, they have to have the appropriate conditions to develop.... With attention, you have to have as much as possible a non-stressed parenting environment, and in this society this is available less and less."

"Why?" I ask. "What's going on?"

Not only is society as a whole becoming "time-deprived," Maté tells me, "People's time is more and more fragmented. People are doing too many things, and parenting tasks just become one more thing that you try and do in a day, and that can't help but affect how our children develop."

I'm reminded of research I've read. According to a British study, parents spend a mere eight minutes a day talking with their children. American parents spend an average of 5.5 minutes a day (38.5 minutes in a week) in "meaningful conversation" with their kids, and by 1990 were already spending 40 per cent less time with their children than they did in 1965. I wonder about these effects; kids, after all, are pretty resilient, pretty adaptive.

"It's because of attunement," says Maté. "For the parent to be attuned to the infant, it's not enough that he or she knows the child. They have to be non-distracted and non-stressed, and have a focus of attention which is the child. It's this attunement that emotional regulation requires. In other words, for the child to learn to develop self-regulation they have to be in attuned contact with a self-regulated adult."

This doesn't just require time as units, I suggest. From my own experience I know that it requires being slowed down enough to be both inwardly and outwardly attuned, mind and body. That way, I could be in sync with my son. I could sense when he was upset. I could pick him up and soothe him until he settled again and was calm.

Maté agrees. "The parent has to be able to sense the child's rhythms and respond to him. It is as we are responded to that we learn to be in tune with other people's rhythms. Where that is interfered with we don't learn that, and so people with ADD, for example, have significant social problems simply because they can't pick up social cues and they don't know how to respond to other people's rhythms. In this supposedly time-saving world, people have less and less time, less and less sense of being centred and feeling focussed, and this is being passed on."

What's being passed on is stress. Stressed parents tend to have children who are stressed and who grow up much less able to cope with stress. One of the many startling revelations in a *Globe and Mail* series on children was that one of the prime signals of future poverty and struggle is kids not being "school ready" when they attend their first classes. Current research reveals that although nearly 30 per cent of vulnerable kids are still from poor households, "the largest number of vulnerable children were spread through the middle class," notably in upscale communities of working professionals.

It's not so much that kids who develop ADD are anxious about time, it's that they have an underdeveloped sense of time as continuities and as rhythms moving from crescendo to diminuendo over time—the stuff of self-regulation and social integration.

Maté explains, "There is no sense of time for the infant. There is only the present. There's no sense of the passage of time, because that's a developmental thing."

I think of what psychologists have written about how children   169
by the age of twelve develop what has been called "a feeling for narrative," that is, a sense of history and of themselves as part of it. "So, this time intelligence doesn't fully kick in?" I ask.

"No," he answers. "The ADD adult walks around in some respects with the time sense of a young child.... There is only the present."

"Does this mean," I ask, "that there are people who, behind the mask of adult calm, are inwardly bouncing up and down like kids on the edge of a tantrum?" I think of road rage, checkout rage. I imagine people in elected office like this, heads of state and of corporations.

"Yes," he tells me. "People who have no sense that the present moment is connected to the future. There are only two things: now or not-now. And if it's not now, it seems like an eternity. That is the time sense of the young child, and for the ADD adult it is very much like that.

"There's a tremendous frustration," he tells me. "There is also a tremendous immaturity in the sense that you really feel that you can't do without things that you actually really could do without.... Whatever you want, you have to have it right away.... "

"Sounds like a boon to the consumer economy," I say.

"Yes," he says, and confesses to a vast CD collection and some equally large credit-card bills indulging his compulsion.

We chat about consumer culture, being a participant of which, it seems, almost requires complying with its compulsive pace, not only of acquisition but of upgrades. It seems that kids are being conscripted early these days, as popular culture grabs for their attention now, now, now, and forever. Apparently, North American kids can respond to a logo before they can recognize their own name. In Maté's opinion, today's consumer culture is largely responsible for widespread pseudo-ADD in which people exhibit the symptoms of easily distracted attention, lack of focus and sustained concentration without necessarily fitting the full clinical profile.

TECHNOLOGY, INACTIVITY AND THE GROWTH OF PSEUDO-ADD

It's a truism that kids spend more time watching television than they do at school, vastly more time in front of the tube than at their homework. By ages three to five, which is considered prime time for the brain's cognitive and language development, North American children are thought to be watching an average of twenty-eight hours of television a week. *Sesame Street,* artfully brought to us by numbers and letters of the alphabet, had a lot to do with legitimizing early-childhood viewing as somehow "educational."

Although the viewing hours have dipped slightly in recent years, especially for older children, that's largely because channel surfing has simply been replaced by Web surfing, on-line games and hand-held ones like Nintendo. It's quickly becoming a second truism that kids spend a lot of time on-line. By 2000, seven- to fourteen-year-olds in Canada were spending an average of 3.8 hours a week on-line, surfing just for the fun of it, playing and downloading games or music, e-mailing and text-messaging friends, visiting chat rooms and, at the bottom of this list, doing homework. Forty-one per cent combined their on-line time with other activities such as talking on the phone, eating or watching television.

With technology, family life is becoming more and more a case of multiple solitudes. More parents are working evening and split shifts, moving past each other in sequential custody and care of the kids, with little time for conversation and bonding. Harried professionals and managers are cramming for exams or working unpaid overtime at home. As a result, they can hardly resist leaving their kids in front of the television or the computer. (Teachers I meet corroborate this: "The parents can't follow through at home. They sign things that aren't done," one tells me. "That's why tracking sheets fall apart," another adds. "They don't follow through on comments that teachers make." Another teacher continues: "In our community, I think we're very much two-income families or single parents or—what's the word?—reconstituted or blended families. But I really get a sense that parents aren't there for the children in the same way that my parents were there for me, and so they compensate in ways that aren't that helpful. It's whatever can be done quickly, but not if it entails any kind of engagement or commitment. I've had parents tell me, 'You do something about the homework at school. I don't have time at home.'")

Family mealtime often means the drive-through at a fast-food restaurant or the takeout section of the supermarket, on the way home from the pool, the rink or the day care. (One in ten meals is now apparently eaten in the car. It's called "carcooning." To support it, the food industry has added a new line of fast foods, including hand-held breakfast cereals, "meat sticks" and yoghurt in squeezable tubes.) Partly as a result an estimated 37 per cent of children between the ages of two and eleven in Canada are overweight, and half of them are considered obese. Aside from large portions of fattening food, kids are eating more because they are stressed and don't feel in control of their lives. In a British survey of 4,000 eleven-year-olds, the most stressed kids were eating more than twice as much as their less-anxious classmates at meals, then supplementing this food with frequent snacking. As well, a British study found that three-year-olds were sedentary 81 per cent of their waking hours, five-year-olds, 78 per cent of the time. If they're not glued to the TV or computer game, they're strapped into car seats by parents afraid to leave their kids alone at home, or outside. Very simply, kids are not getting out to play.

Time for play is essential for a healthy sense of ourselves, being engaged in the world and knowing who we are. Now, however, playtime, especially shared playtime, is no longer the norm but the exception to how kids spend their time in many, particularly middle-class, communities. In fact, some of the centuries-old games that older children passed down to the younger ones, including double-dutch skipping rhymes and Mother May I are being forgotten, and what author Merilyn Simonds calls the "underworld of children's culture" is being lost.

"You have to teach the kids recess" now, according to one school administrator. "I see kids all the time in Grade 5 who get here and they don't know how to play," a private-school headmaster told a *Globe and Mail* reporter, and blamed it on "too much pressure on children."

A lot of kids also are being overprogrammed by parents anxious for them to get into the right schools, learn the right skills to get into the "good jobs" side of the "good" jobs/"bad" jobs divide. I recall talking about this with Ursula Franklin, University of Toronto professor emeritus, author of *The Real World of Technology* and an active grandmother. "Remember that old saying where you ask a kid: 'Where do you go?' 'Out.' 'What will you do?' 'Nothing.'" she said. "This sort of wondering about what will occur… the unprogrammable, which is very much a part of life, is being more and more squeezed out by programmed, scheduled activities. In my opinion, this is where creativity lies. That's where dreams are dreamt. We are deprived, particularly the young are, of unscheduled, unprogrammed intervals of time. This may have a very severe effect on the freedom of their imagination."

Are children's imaginations growing in other ways, fed by environments other than the increasingly barren backyard? I wonder. A number of computer games are challenging and seemingly constructive. Kids spend hours, often together, playing "The Sims," a series of simulated-world games in which players can not only build homes and cities but populate them with characters that come to life through multiple-choice scripts and non-verbal actions like "hug" or "tickle." Still, it's a little unsettling that the scope for play and expression involved is limited to multiple-choice options predetermined by the game's inventors and that the entire playground is fictional, a simulation in which consequences are immaterial, and so are human relationships.

I worry, too, at other conditioning effects of these games, especially in light of the "militarized masculinity" of mastery, control and killing on which so many computer and video games turn. Quite apart from its possible link to renewed sexism and even misogyny, neurologists studying the body chemistry of people playing hand-held video games have found such a tight identification between players and their on-screen characters that, as the violence and frenetic action escalates, not only do their blood pressure, heart rate and adrenalin levels go up but the part of their brains associated with physically preparing to fight is also activated.

I can sense how pseudo-ADD can be at work as, research has found, children who play these games come to believe that "faster is better" and that peril is nothing. Equally worrisome, video games' high-speed driving simulators actually deactivate brain areas that sense risk and counsel caution, even empathy. Other studies drive home the particularly pernicious effects of violent video games. In one, after playing a lot of violent video games, teenagers with normally non-aggressive personalities became almost ten times as likely to get into a physical fight as were teens who didn't play the games, a rate that topped the fighting propensity among already aggressive teens who didn't play the games. Other studies continue to document a link between schoolchildren watching violence on TV and engaging in violence such as schoolyard bullying in real life. Another study has identified a wholly new problem—the rise of "cyber bullying" in the form of e-mails or short text messages sent by cellphone—attributed, in part, to the anonymity of this virtual communication and to being distanced from the real-life consequences of one's actions.

These problems aren't new. However, the conditioning and deconditioning effects of these simulated-play environments could approach a critical mass. This distanced, fragmented, moment-centred, ego-centred way of functioning in the fast, immaterial world behind the screen could become the basic template for how people engage in real life—as the spillover effects of violent video games suggest. Traditionally we've believed that environments are separate from us, and that people adapt somewhat deterministically to them as given, in a kind of do-or-die struggle for survival. But recent research in developmental biology suggests that we co-create our environments for living, in a dialogue of mutual accommodation. As Harvard biology professor Richard Lewontin points out, "There is no environment without an organism. Organisms do not experience environments. They create them... by their own activities." And so, much as robins now use shredded plastic along with the usual grass, twigs and mud to make their nests and perch them on light fixtures as much as in trees, young people assemble an environment in which they feel nested by wearing the latest designer clothes, mixing and mashing digital audio-tape bits and updating the signature dial tones on their cellphones.

Although it's likely that music and fashion will continue to evolve and the world of interactive graphics will be cleaned up, the larger question for me is how to ensure that the forces that underlie these will teach young people to engage in the world of face-to-face reality not just the virtual realities on their computer games, personal digital assistants or cellphones. It's not just that kids aren't going outside much to play any more, it's also that they're staying inside the one world where they are allowed to roam. This micro-environment that is instant, changing constantly and therefore unavoidably ephemeral may feel most like home. The zone of flows becomes their comfort zone, along with its sometimes compulsive pace.

ADDICTIVE ENVIRONMENTS AND THE CHALLENGES OF REAL LIFE

Among kids raised on a diet of television and video games, addiction is a growing concern. Essentially, the fast pace of change and instant sensory stimulation effectively condition the brain toward external stimulation and direction, rather than internal self-regulation and focus. This can happen in part because our brains are almost hard-wired to be alert for sudden changes in the environment around us, because in the primordial past these might have signalled danger. So kids—many of whom have already been diagnosed with ADD—are consuming alcohol, cigarettes and other drugs at a younger and younger age, and they are becoming problem gamblers, too. It's telling that youth addicted to gambling aren't playing to win. What they're addicted to is the adrenalin rush of the gambling act itself, to the "arousal level" and the sheer "excitement" of the play. In other words, they're addicted to the "jolts per minute" that link so many elements of popular culture with pseudo-ADD. They're used to having their senses aroused, not to the larger end of being engaged in a tactile, full-sensory dialogue or adventure with a friend, but just for the rush. The focus is on precisely what so much of television and popular culture generally delivers: sensory stimulation (jolts) with neither the time nor the opportunity for any follow-through or involvement.

This absence of opportunity to engage the attention once the senses have been turned on means that "the impulse [to act] has no outlet," which researchers have linked to hyperactivity, frustration and irritability in children. At the same time, it leaves children primed for constant stimulation without sustained involvement as a dominant form of being in the world, almost inculcating a social pulse more attuned to the beat of the games than to the rhythms of real-life conversation and relationships. In fact, researchers are also finding that addiction is linked to dissociation, to the chance to exit one's regular identity (and any problems associated with it) and revel in "altered egos." It's "the ultimate escape," they say, with time passing quickly for as long as people are playing the machines.

Tellingly, therapeutic research on kids with ADD finds that they perform well, and are able to concentrate and stay focussed, as long as they're in a game-like learning environment with lots of instant rewards, but only for as long as the stimulation and rewards are provided. This sounds like a perfect set-up for an adulthood of compulsive work and equally compulsive consumption, if more obvious addictions haven't got to them beforehand. In turn, this might be great for the new economy, with its nanosecond turnover rates, requiring matching rates of production and consumption. But it leaves people with ADD rather ill-equipped for the real work of life, that is, creating a meaningful existence, establishing lasting relationships of love and friendship and contributing to social institutions at work and to local communities at home.

Throughout the era of the social-welfare state, we have relied on our schools to fill in the gaps. Schools have provided breakfast programs for students in impoverished areas, teachers have provided encouragement and a listening ear for kids with special needs or problems. It's long been assumed that teachers can help young people grow up and become good citizens, but the public-school system has been radically changed by cutbacks and restructuring. Not only are teachers among the most time-squeezed professionals these days, the new rules and regulations being introduced to govern and direct education are taking their attention away from the kids.

THE CHANGING NATURE OF TEACHING AND LEARNING RELATIONSHIPS

When I talked with some Toronto- and Ottawa-area teachers, I learned that their top concern is not overcrowded classrooms but over-fragmented and over-scrutinized classrooms. Education has become both highly individualized and geared to broad, almost global standards of performance. Individual education plans (IEPS) were once associated with a minority of kids with special needs. Now they are becoming the norm as children are being passed along from grade to grade according to their age rather than their ability. As one result, teachers don't address the class as a group so much any more. Nor do they assess the children with broad benchmarks of development in ability to communicate, to grasp and manipulate symbols, to get involved in class and get along with others. In Ontario, where curriculum restructuring was a major priority of the Conservative government in the 1990s, teachers have to assess each child's performance according to a detailed set of "overall" and "specific" objectives, starting in Grade 1. (In the Grade 1 curriculum, for example, there are seven overall objectives laid out in the category of "reading" alone. Within that, there are fifteen specific expectations, grouped variously under the headings

"understanding of form and style," "knowledge of language structures" and "use of conventions" such as punctuation.)

By and large the teachers support the new curriculum, in principle. It's an intelligent articulation of an ideal path to learning.
The problems arise in trying to make it work in the real world of
cutbacks and of six-year-olds who are still sometimes wetting their
pants, or still learning English or French, or are having trouble
staying awake long enough to take in what's being read or taught.
Plus, the effort required to prepare the IEPs and then to track kids'
performance through all these detailed fields on the computer's
reporting screen takes up so much of their time!

"Take one strand (one 'specific expectation') in math for Grade
1," one teacher tells me. "They have to count by twos to 100. [Test-
ing them,] I can't just fake it. I have to sit with each child and go
through it all to make sure that the data are right."

Her colleague adds: "And you start assessing them as quickly as
possible, and then very often we feel inadequate because we loathe
to be teaching to the report card. A lot of us hate that. It's unnatu-
ral. You want to teach from where the children are. That's where
the best teaching happens."

This is their main quarrel with the current educational reforms: The quantifiable stuff of test results is eclipsing the unquantifiable stuff of social engagement. From the teachers I talked
to, their gut instincts reject this.

"I will let the curriculum go before I would let that go," one
tells me, "because otherwise I'm not going to get through the curriculum anyway, because if we don't have a relationship that goes
beyond just 'I'm going to teach you this and you're going to respond,' then we're not going to get anywhere anyway, so you have
to spend that time."

Another adds: "We all know that unless you have that connection, that relationship, you're not going to be very effective as a
teacher, regardless of the student's abilities, or 'modalities' [that is,

how they learn visually, kinesthetically or otherwise]. You have to have that relationship or rapport as well."

The first teacher concurs, then continues, "But then I would say we might not be the best models at times, because we are constantly on the go, and we've already discussed that we feel totally stressed and pressured and fragmented to do so many things. So, again, we ourselves have to really force ourselves to work with the children over time, and let them have time. It's not easy. I think they're getting mixed messages, really."

The teachers are also giving mixed messages to each other. Although they talk about taking their time with kids, they no longer have time for themselves. In both cities, they commented on how empty and silent what has traditionally been called the teachers' common room has become. The medium is the message, and the medium that is the social environment of schools these days has become so scattered into specific-objective lesson plans that the culture of education is changing. Schools are losing their integrity as institutional wholes. Partly too, though, this is from outside accountability audits and the pressure for quantifiable outcomes per individual student and class. In the classroom, this translates as a focus on the detailed objectives for each strand of each subject being taught. The objective is to get through each task and related test, in the time allotted per strand.

"So you get the chopping up of the child, but you get the chopping up of the teachers as well," one teacher says.

"We don't really teach mastery any more," another adds.

"Oh," I say, surprised, mastery and excellence being such buzzwords of our times. "What do you teach toward, then?"

"Moving on to the next piece. You don't really go into anything in depth any more and keep going over it so you feel that the children master it."

I ask about that wonderful "aha" moment, the thrill of the lights going on, the door of understanding finally being flung wide open.

"A lot of kids aren't saying 'aha' any more," she says.

"What are they saying instead?"

"'I don't know,'" she tells me.

One of her colleagues nods, and adds: "I think on top of ADD we are seeing many more depressed kids." (Sure enough, a growing number of children are taking antidepressants, for both anxiety and depression. Children as young as eleven to fourteen are also starting to be sleep-deprived, which is another trigger for depression, as they stay up late doing homework on the computer and e-mailing friends.)

Meanwhile, teachers are increasingly focussed on teaching to the test, because schools are now being ranked according to grade-point-average results.

"We're all median-obsessed," one tells me, with a smile that is part humour, part apology. "What if I've got a whole bunch of Grade 12 students who just won't do the work because they really don't feel committed and their marks are low? Then I start worrying. I remember someone saying to me, 'You can't have marks that low' [because it would make the school look bad and him as a teacher look bad]. So what is my choice? None of the choices are good."

I ask him to take me through the choices.

"Well, teaching to the test is one thing. In Grade 10 now, we're already teaching to the [high-school graduating] exam. So you can forget about all those wonderful ways of writing essays, or writing at all, because the only one that is going to be tested at the end, the one the money is on, is the five-paragraph essay called expository writing, which starts with a thesis, gives three good reasons to support that thesis, with specific evidence from the primary text to support the reasons that support the thesis. You know what I'm saying?"

I nod. "Formulaic writing."

He continues. "And it's going to be marked on a five-category grid. That's what's dominating everything. So, just get in the harness and do that.... That's a nasty way to beat the system, but that's the way we're all going.

"The other thing is, we've been using computer programs for marks and, well, actually I can look at my marks and I can toggle one little thing and, oh, the class average is up to 60 per cent. I reweighted something, and that solved the problem. That's another choice.... There's all this pressure."

"Pressure?" I probe. (You mean to 'cheat'? I add inwardly.)

"To maintain a class average that is acceptable," he says carefully, meaning, delivering what is acceptable to the outside scrutineers of quantifiable performance.

Looking ahead, many teachers worry about a larger political agenda: an increasingly polarized, possibly two-tier educational system in which parents use government-issue vouchers to enroll their kids in the schools producing the best grade point average, where kids are groomed to outperform the competition in the global economy, by teachers groomed to help their school outperform its competitors in the market for the best students. At that point, the political struggle against schools being run as businesses rather than publicly, for engaged citizenship and self-development, becomes moot. Meanwhile, there's the larger ethical issue of statistics and spin-doctoring with little or no accountability to the core human values of education and culture.

In my opinion, schools need to provide a counter-environment in a society where so many other institutions of the social landscape are wired into the global and instant hypermedia environment. Kids (all kids) need a chance to cultivate a healthy sense of themselves as articulate agents of their own lives in society, through the time-honoured traditions of conversation and discussion in a classroom. They need this grounded environment more than ever today, as they prepare to spend more and more of their lives on-line. As educational psychologist Dr. Lillian Katz put it, kids growing up in a high-tech world need "a low-tech, high-touch environment"—where, I would add, the pace is slow enough for kids to actually feel that touch and be motivated by it.

I'm ambushed by a memory of a little girl with freckles and a frantic-to-please smile. Small for her age, she's wearing a navy blue tunic, white blouse and knee-high navy blue socks as she walks into Mr. Bennett's Grade 7 class. She's been moved into this class because her mother has insisted that no child of hers is either slow or retarded. Nonetheless, at the end of Grade 6 it was clear to all of her teachers that she couldn't read, not to the point of grasping the sense of the text. It's tough being in a new class, among the smart kids, a lot of them smart-alecky and given to calling her names. Still, one girl befriends her, and she develops a crush on her teacher. Mr. Bennett is a roly-poly Englishman with a big red moustache and a wonderful chortling laugh. He beams down at her. He encourages her as she stumbles along, explaining what she can't understand. Slowly the glaze recedes from her eyes, and by the end of the year she can read.

I blink back tears, and think: I finally learned to really read that year, because somebody took the time to pay attention to me.

Eight

DRAWING STUDENTS INTO
SOCIETY'S CONVERSATIONS

"We empty of their humanity those to whom we deny speech."
GEORGE STEINER, "A Future Literacy"

"I wanted the McGill brand name on my resume."
EVE YANG, Maclean's

THERE'S A MOMENT I wait for in the seminar course
I teach every year at Carleton University: the moment when the students begin to engage with the
ideas and knowledge I bring to them every week. They
start to incorporate this learning into their own base of
knowledge and opinion and interpret it back to me as
something wholly fresh, wholly authentic and original,
as they understand it. This moment usually occurs
around the time when the students first remember each
other's names, a moment for which I also watch and
wait as September moves into October. The roots of relationships have begun to take hold. The students are
beginning to become engaged, with me, with each other.
And I always rejoice when it happens, because to me
that's what teaching, at least in the liberal-arts and humanities tradition, is all about: encouraging young people to think for themselves, to speak their own minds

and listen to each other and to realize that what they think and say matters.

People who champion this more "political" component of post-secondary education associate it with civic leadership and involvement. They emphasize the critical thinking skills, the communication skills developed not only in seminars but in informal chats during office hours or in the hallway. I'd go further and say that this practice of becoming engaged, through thoughtful dialogue with a professor and fellow students in a seminar, can cultivate the confidence to become an "implicated participant" in society—that is, a participant with a sense of something to give, and with this, some feeling for and commitment to the common good. Out of our shared time together in that room, we create a small crucible of democratic practice, where everyone has the right and the responsibility to participate. It's literacy writ large: the ability to knead the dough of experience, knowledge and understanding and, from this, to articulate yourself into being as present and accounted for in society, having a stake in the larger issues and having something to say as well. Very simply, it's learning how to read and interpret the world, as Brazilian educationalist Paolo Freire viewed it, and respond.

Learning theorist David Geoffrey Smith defines literacy as including not only reading and writing but also speaking and listening, requiring not only skills but the sensibilities of attunement and attention. When it succeeds, students share their thoughts and nurture what is essentially a gift economy of shared knowledge creation.

As the 1980s gave way to the '90s, however, I noticed that the moment of engagement I always wait for was happening later and later in the course. I had a sense that students weren't giving their undivided attention, so one day I asked for a show of hands and sure enough, all but two out of the twenty students that year had jobs. How many were working fifteen hours? I wondered. All eighteen hands went up. "Thirty hours?" I asked, knowing that

184

summer-job earnings have actually declined since the early 1980s while undergraduate fees have quadrupled since then to more than $4,000 a year. A significant minority of my students raised their hands.

Today's students hardly have time even to complete their core class readings, let alone to follow a tangent of interest or to go for coffee together to discuss their findings. Twenty years ago, when my students did group presentations they routinely met to thrash out what became a collaborative kickoff to the class discussion. Now, I find, they tend to e-mail each other or use their cellphones to divide the topic into strands, which they separately research and present as individuals. Their time has become so squeezed that my students have become almost too focussed, especially, as one academic at McMaster University in Hamilton, Ontario, commented, on what's related to the labour market. Increasingly, I find, students will put time into something only if it will directly affect their mark. Some have become quite competitive, too, checking to make sure I assign individual marks for group presentations, shielding the notes they're taking from the person sitting beside them.

Lack of preparation time is impeding the discussion, but so is something else. Though they take lots of notes and mark their readings with multicoloured pens, my students no longer seem to retain much of what they've learned or read, nor do they recall it at appropriate moments. In fact, at least one study has marked as a new trend among young people aged twenty to thirty-five a worrisome inability to remember. Partly it's blamed on an over-reliance on computers (plus on-line information) as an "external memory." Partly it's thought that young people are often too distracted when they're reading to absorb much in the first place. And partly it's thought to be the result of information overload. I wonder if it's also due to an overdeveloped sense of "learning" as assembling and downloading ready-made knowledge bits, and not enough of a sense of the participative element. It's a truism that you remember best what you work with. Memory and dialogue go together, including through the inner dialogue of thought. What does it mean for society at large when the brightest of our next generation don't develop an ability—and an appetite—to engage in this dialogue?

DIALOGUE AS CIVIC ENGAGEMENT

In my bit of the ivory tower at Carleton University, I worry about being able to cultivate these future "wise citizens." As Canadian communication theorist Harold Innis wrote: "Culture is concerned with the capacity of the individual to appraise problems in terms of space and time and with enabling him (or her) to take the proper steps at the right time." In other words, we want citizens with the capacity to think things through with enough depth and confidence that they can act, and act appropriately, when the occasion warrants it. Memory, or at least the retention of information, is crucial to this critical thinking, in both its aspects: the broad range of personal experience, chance observations and overheard chat, plus the deeper wells of accumulated and composted knowledge, wise nuggets and one-liners against which we test new information, sensing contradictions and resisting accepting the latest idea or news flash simply because it is new. Both types of memory come together in critical thinking and in the verbal practice of this thinking in seminar dialogue, where serendipitous cross-associations and juxtapositions from side comments can spark new insights and students learn to speak their minds, in their own voice. (Again and again educators stress this point, about people finding their own voice and, in the process, coming alive to the notion that they are participants in the workings of their society, not just consumers, not just bystanders.) As students learn to use their voices responsibly, which includes backing up their opinions and listening attentively to others in the seminar class, they cultivate the social habits of civic engagement, including those of civility.

Classroom dialogue prepares students to ask and engage in the larger democratic questions wherever they go and to resist becoming technocratic decision makers—that is, people who rely on ideas, facts and techniques for analyzing them and see knowledge as the stuff of information systems, separated from the larger living context of society. This "objective," data-based way of knowing, and its rise to a near-monopoly position in public discourse and decision-making, has a long history, which I'm too ignorant to even attempt to explain (or smart enough not to try.) However, a couple of points are worth mentioning, as they apply to what I'm saying about dialogue. First, this is knowledge as represented reality, organized around rules, standards and abstract concepts, that is not only stored and produced outside ourselves (think of books with print runs in the hundreds of thousands or databases that can be copied indefinitely, in nanoseconds) but is of a scale and scope—and growing at such an exponential rate in the new economy—that dwarfs any insights we might "produce" ourselves with our own spoken or written words. More importantly, the dwarfing operates at a deeper level because the knowledge that's derived from outside ourselves, even outside the realm of direct experience and its stories, is also valued more highly because it is considered reliable, whereas personal experience and perception are not.

This idea of knowing based on doubting the veracity of our senses dates to the sixteenth century and Italian astronomer Galileo Galilei's telescope. If we humans could be so wrong about the sun orbiting Earth, which the telescope put the lie to, modern "rational" thinking asked, how else could we be deluding ourselves? The dethroning of Earth as the centre of the universe was accompanied, then, by a sort of parallel dethronement of the body and bodily senses as an authoritative basis for the truth and a shift toward ever-greater reliance on science, its instruments and its rational thinkers, with their theories and supporting data and texts as the source of this information.

Dialogue, including that in seminar classes, is a way of rooting this inert knowledge back into the realm of living bodies, the life of society and the living earth. It's a bridge, mitigating the distancing

187

influences of scale, scope and speed; it's also a medium for making this data (and, by implication, the technocrats associated with its production) responsive and accountable to the broader population and the realities of society. As Boston University professor Alan Wolfe argues in *The Human Difference*, it's not our ability to process data according to set rules that marks us as human. It's our ability to *interpret* knowledge and information, improvising new rules as circumstances change.

Dialogue in the classroom, even and perhaps especially among technocrats-in-training, can help students develop these bridge-making, interpretive skills, plus the sense that these are essential for a healthy culture and society. In fact, as they rehearse this, along with discussion-leader teachers like me, they might be modelling new understandings of our universe and ourselves as participants in it. In his classic *The Dialogic Imagination*, Russian literary critic Mikhail Bakhtin argues that dialogue is the cultural rendering of relativity theory. The key, he feels, is the new focus on "relations among things" as experienced in time. Nothing exists in isolation, as pure time or pure space. Stories model this, he says, with their inner dynamic of dialogue between elements and between the text and the reader or listener.

ROOTING LANGUAGE IN THE BODY AND IN TIME

"I do not *use* language; I *am* language," Montreal poet and educator David Solway wrote in a savage critique of the borderline illiteracy apparent among post-secondary students these days, and his equally savage criticism of the rote remedial reading that is so often prescribed in response. For him, the problem is not that young people are unable to read or write, or cut and paste, but that they suffer from what he calls "aphasia," an inability to articulate themselves as social and historical subjects. Eschewing the language of attention deficit disorder, he has coined his own term, "chronosectomy," which he describes as a kind of lobotomy in that part of the brain that senses time as personal and shared continuity.

When I visited him in his lovely, leaf-dappled home in the Hudson hills outside Montreal, I asked him to elaborate on his concerns.

"As we've entered this cybernetic milieu, we have externalized functions that we had always, up to this point, understood as being internal functions. The inner sensorium for analyzing, parsing and experiencing time has been outered [placed outside ourselves] into a kind of technological surround so that our memories... " He breaks off.

"I had understood," he continues, "that what my students were suffering from was a deficiency of memory: they did not live in time in the two memorial senses of historical relatedness and personal continuity. A sense of process, of movement, of implication, of engagement is dependent upon a subject that lives in time, and that is internally and temporally structured."

For some reason, Solway tells me, young people today aren't getting this idea of being part of, even immersed or embedded in, a continuum of time, and as a result they seem to be simultaneously moving at the speed of light and going nowhere.

"You move so fast that you don't have time to absorb what is happening within you, and to make it part of your own sensibility," he says.

"It's the great video arcade of time.... Everything glittering in the moment. But the glitter is encapsulated in the immediate, in what [poet William Butler] Yeats called the 'glance.' Everything is a function of glances, as opposed to the gaze, which lingers. He was worried even then that we were moving out of the culture of the gaze, where we linger on things that exist around us because they are inherently beautiful and valuable in themselves, into the culture of the glance, in which we only perceive instantly and then forget as we move to the next glance."

For Solway, remediation requires bringing these young people back to their bodies and a sense of themselves as subjects capable of speaking and acting through an engagement in life beyond the fragmented moments of chit-chat. No matter what subject he's teaching, he sees his primary job as "re-enacting the very function of language itself" and, by modelling that to the students, encouraging them to do the same. His remedy is simply to get them talking and discussing, learning to speak themselves into existence as full-bodied participants in the event called life, learning to articulate time as they experience it directly and as they share it with others in class. To do this, he says, the entire educational milieu needs to be preserved as a "slow zone," where students and the adults who are committed to helping educate them (from the Latin *educere,* meaning "to lead out") can "take time for time." That is, they can bring established knowledge and make it fresh in their discussions.

Solway is adamant: "When language succumbs, perception and thought are equally damaged."

BACK IN MY OFFICE, preparing for that day's class, I thought about our conversation as I worked hard to resist all the pressures (the blinking light of my unanswered voice mail, the discreet beeps alerting me to incoming e-mails, the vaults of information at my disposal on the Internet) that would take me away from this task. As I reviewed my core notes and marked passages in favourite texts, I sensed the key phrases coming off the page. I felt them becoming three-dimensional. They came to life as I felt my passion, or my cynicism and sense of irony given whatever's been going on in society lately, reanimate the written words. By the time I was ready to meet my students I was pulsing with energy, as alive to the material as I want my students to be. Because, like David Solway, I'm holding out for a more democratic and cultural definition of literacy that encourages people to read the situation and not just the instructions, and to speak with their own voices, not to become the hands and mouthpieces of corporate texts.

I don't just want them practising for a "best-practices" performance in a virtual version of the world, and I don't want to model that to the students, either, by compressing the prep time I need for full participation in the here and now. That's why I rethink what I want to say every time I teach this course. That's why I cap the enrollment. I consciously cultivate a micro-environment that is scaled for the particulars of different people and is slowed down so that it can serve as an antidote to the conditioning effects of compressed abstractions and speed-up. I want students to put themselves into the picture, to sense themselves as part of society's larger story and even helping to articulate it. So I model that in the medium of my class. But, as I'm equally aware, it's getting harder and harder to do this on university campuses today.

SPACE-TIME COMPRESSION AND THE WIRED UNIVERSITY

On the surface, nothing much has changed on a typical university campus since I was a student in the 1960s and part of efforts to democratize post-secondary education and its curriculum. The face the institution presents to the world, in advertisements and in the annual report cards ranking universities across the country, still features an ivy-covered stone building or two and gaggles of young people hanging around on its steps, talking and laughing. Wandering around the Carleton University campus, the acute observer will notice that there are lots of Coke machines—only Coke machines, it seems—and suspect that Coca-Cola has the exclusive supply contract on campus. A stickler for detail might worry at what this implies about the university's commitment to freedom of expression in this space supposedly consecrated to that purpose, though increasingly dependent on corporate funding for survival.

The biggest changes in universities have occurred behind the scenes, epitomized best in the wiring that runs behind the walls, above the ceiling panels and under the floor beneath academics' feet. This wiring has turned the university into both a local digital network and a fast subject in the on-line world at large. Some of the impetus came not from academics themselves but from administrators. Drastic funding cutbacks in the 1980s forced universities to be more conscious of how their money was spent. Questions were asked. Standardized reporting and accountability procedures were implemented. As funding cutbacks became business as usual, ad hoc survival strategies jelled into a more centralized, businesslike decision-making process on campus, with a lot of on-line consultation and reporting to facilitate and support it. These computer connections extended beyond the campus, too, as schools reached out for corporate donations to offset the loss from public coffers and individual departments sought matched funding for research.

While the arts, social sciences and humanities departments shrank under the relentless effects of funding cuts and downsizing, some of the newer areas—particularly the professional schools like business and engineering—grew, in part from corporate funding and applied research and consulting. Naturally enough, corporations demanded their kind of accountability in return—not so much overtly in harmonizing research and teaching priorities with business priorities, but indirectly through standardized and quantifiable performance indicators that could be compared from one university to the next, for instance in the annual university-rating reports in *Maclean's* and the *Globe and Mail*.

Critics have questioned the "corporatization" of universities, though, by and large, academics aren't worried. They've generally embraced the changes (even sociology professors dress more smartly these days, leaving their Birkenstocks under their desks!) since, being globally connected twenty-four hours a day, seven days a week, they can do collaborative research with people across the country and around the world, not only in other universities but in not-for-profit institutions and as consultants in the private sector. Being "digital" also gives academics greater exposure for their work, and, with on-line journals multiplying the rate at which new publications are launched, they can build their curriculum vitae for annual performance review, tenure and promotion. As well, connectivity has eased the burden of cutbacks in support staff; academics now file reports and marks on-line. They

192

do committee work on-line. They download new administration policies from servers on their network. They post course outlines and class notes to Web sites, although many worry that doing this will result in even fewer students bothering to attend class. Part of and even entire courses are taught on-line, and sometimes the lectures are broadcast on local cable television. And if it all gets to be too much, academics can work from home on-line, deal with students and colleagues over the Internet, hide out on-line. At least, that seems to be the picture emerging from an academic-time survey I co-initiated with Janice Newson, a social scientist at York University in Toronto.

However, not only are academics compromising their health (through stress, insomnia and attendant problems with memory and concentration), they're compromising their intellect in ways that I fear must affect the kind of role model they provide for students. For example, an overwhelming majority of academics in the study said they no longer read as broadly and interdisciplinarily as they used to and as they'd like. They were not reading as deeply and reflectively, either; the majority were reading "more narrowly" than they used to or than they'd like. They were also tending to only scan journal articles and books, mining them for selected bits of information that they could use in their own work. Hardly anyone had time even to read what colleagues have written.

I wonder if it's a form of self-censorship, this silencing of academics' authorial voices and range, or a species of self-forgetfulness and neglect, which would be ironic given how immersed academics are in the world's knowledge networks. Yet they're hardly free to speak or even know their own minds if they lack the time and space to interpret all that information for themselves. And that appears to be the problem.

It seems that while universities are branding their identities into ever-clearer mission statements and logos, they're also disintegrating as organic wholes, as a commons of shared ideas and, on a more pedestrian level, simply as shared space and time. In follow-up interviews with the academics who participated in the study, they talked about the silencing of dialogue between them even at faculty meetings, where, as a professor at Dalhousie University in Halifax, Nova Scotia, said, "it's like being with quasi strangers" because that's the only time they see each other.

They also repeatedly stressed the importance of being present with the students not just in class but around the campus. They talked about being articulate enough to make their own ideas and research come alive to the students as real and relevant and, almost more essentially, of helping students gain a sense of their own relevance by listening to them, giving them a chance to talk and to find their own voice. Essentially, they were suggesting that they acted in the age-old teaching tradition of mentoring, which in the prototype (Homer's *Odyssey*) features the Greek goddess of wisdom, Athena; dressed as the human Mentor, she directs and encourages Telemachus in his quest for knowledge and understanding. This element of teaching and learning, especially couched in apprenticeships, is perhaps the longest-lived form of schooling in the world and is practised in First Nations societies, in the Muslim and Hindu cultures and in many European universities.

Two of these conversations I had with Canadian academics particularly stand out. One was with a professor of engineering at the University of New Brunswick in Fredericton, who was happy enough to e-mail students as a way of staying in touch and up to date but who wanted to see them in his office to work on design problems, so they could doodle together on a piece of paper.

I ask him whether that's so that he can tell if they're "getting" it, if they can sense that he cares whether or not they do?

He replies, "I think you put your finger on the most important thing. When you respond to their e-mail and say 'Come and see me,' they think you really care.... If they are shy and they come and see me and they find I am approachable, they come again." Plus, he says, "part of my enjoyment of teaching is the one-to-one interaction."

"Why?" I ask.

"Because of the dynamics, the chemistry. I can see if they understand what I am saying, if the point is getting across, if the problem or issue I am discussing is worth elaborating more. I can't see that through e-mail." He laughs, and adds: "I can make a fool of myself face to face and not worry about it." There's far less chance that the tone or the message of what he is saying will be misconstrued.

As well, he tells me, in engineering there's no one perfect answer, so going over a problem with a student may help both of them to look at something in a different way: "In many ways it is trial and error.... If I want to build a structure, defining what I want to build is half the problem. It isn't really an absolute. We say the structure could be very rigid or very flexible. We can solve the problem this way or solve the problem this way. A solution that might be acceptable under these circumstances might not be acceptable environmentally or something. You know what I mean? So you can have more than one solution for the same problem. But how are you going to do that kind of dialogue on-line?"

I'm fascinated by what I've just heard. "You're essentially saying that the medium is the message," I tell him. "You teach engineering so that your students are prepared to deal with the unpredictabilities of life. It's as though you try to model this, to rehearse the fact that the real world is always different, it always depends on the circumstances, the context."

"Yes," he replies. "The problem is open-ended. If it is closed, idealization, then innovation is meaningless." He goes on to talk about how even after a dozen years of teaching things still surprise him; he still doubts and questions.

"And doubt, you can't make that electronic." It's the fecund and often contradiction-filled space-time between idea and reality, theory and practice, and it's brought to life through face-to-face dialogue, yet largely lost in the instant abstractions of interface.

At the University of Regina in Saskatchewan, a professor of film studies made similar points. "I tell my students in every class:

'Here is my number. E-mail me if the wheels have completely fallen off and you need to contact me. Otherwise I want to see your face in my office. I want to get to know who you are.' I think, as you were saying, e-mail can speed everything up so—wow—it's going to be a lot easier for me to answer that student by e-mail. It's going to cut time. No one will have to come in and learn the skills of interacting—so I think on the one hand it is a question of 'we can easily close our doors and turn to our research and not have to be involved,' and on the other hand it is a question for teaching, too.

"It is easier if everyone is anonymous. I've had an e-mail from Marilyn Bell, or whoever she is—but I can't even pick her out from the faces in the class."

"So that makes it easier, if she's without a face, without a physical presence?" I ask.

"It makes it easier to refuse social interaction," she says. "It makes it easier to turn people into things and to just not be creating a humanistic environment."

"Are you thinking that as we move more on-line we are moving toward a more transactional form of communication rather than something deeper?" I wonder, thinking back to what I know of communication and how its first meaning was intimate and fiduciary, focussed on communion.

"Yes," she says. "I think that absolutely. And it's got to do with body, it's got to do with 'Hey, this is a good way to not have to deal as embodied people.' I just see it as part of the whole—this virtual revolution. We are just at the beginning of it, where our bodies are beginning to be sloughed away."

"And that includes being fully present and embodied… "

"… and in dialogue with others," she says, finishing my sentence. We talk about the fact that some 80 per cent of communication is body language. "There's also," she adds, "the student-professor power relationship. We forget about it, but they are just so vulnerable to our sense of knowledge—that we know so much and they don't know anything and part of, I think, teaching at the university level is an emotional relationship as well that says, 'I'm in here to support you as well, as a person, in your development, which goes beyond just whether or not you understand the course.'"

"And you think that requires dialogue?"

"Yes," she says. "I think so. Because to me it has to do with authenticity of person. I'm stuck in the dark ages with a body and I've often felt like that with technology. It is a revolution that I find really dangerous, particularly as a feminist. Women, we have always been the embodied subject, and technology seems to be erasing our presence in the world. I feel technology is damaging our sense of authenticity that way."

I ask if by "authentic" she means the ability to be genuinely original, articulating ideas but filtered through a person's own experience as a historical subject.

She explains, "I don't mean 'originality' so much as authentic in the sense of being an active authorial agent, answerable in word and deed to others. It's about emotional and intellectual answerability to others from one's unique place in existence as an active agent or authorial subject."

I ask if she finds it's harder now for students to feel that they are subjects and fully answerable as they spend more and more time on-line, processing information and consuming it, rather than co-producing it as they express their own thoughts and opinions?

"Definitely," she tells me. "And it's the way younger students, the new generation, go about doing research, with absolutely no depth. The kind of things they will search out on the Internet—switching from this to that, this to that, but they can't seem to… there's just no depth to their reading, often, and there's no depth to their sense of 'What do you do with all this material? How do you focus?'"

"And what do you do to try and counter that?" I ask.

"As I was saying, teaching that 'you have to come in and talk to me. You have to find out what interests me, and I have to find out what interests you in a deeper sense.'

"I had a student in the other day. What was she talking about? It was about writing, and what is she doing when she writes an essay? I started to talk to her from a dialogic perspective. I said, 'You have this field of all this material. Why should you write anything about it? What have you got to say that would in any way be important?' I started to talk a bit about the notion of authorship, and we all have a voice and this fascinating notion that it's all a babble and forms of language and we are all immersed in them all the time but what you have to say is important. And the lights went on for her and she said, 'Well, I never thought about it like that.' And I said, 'That's why you try to begin to question.' In order to do that you have to immerse yourself in the dialogue, in what's gone on, you have to read as widely as possible and then your own perspective on it will begin to emerge, and I think, could I have done that typing to her on e-mail?"

"It could be that we help to kindle that, the very idea of our students having a point of view by our own attentiveness," I suggest, "by our showing an interest in what they have to say."

"And our excitement, too," she adds, excited herself. "I don't know how you can convey that in e-mail language. The way someone's eyes light up. It's interactive."

What's important is the relationships that are actually fostered, that take root in the shared time and space of talking, debating and telling stories together. They root young people into mini-communities and micro-societies, where they don't just talk about combining the thoughts of authorities with thinking for one's self. They do it. In turn, the experience fosters the ability along with a taste for this, at least that is my hope and part of what motivates me in my teaching, because, to me, what's being cultivated is the capacity for responsible talk and action in work and other relationships of their lives, and for accountability. Responsibility starts with ourselves, in the connection between conscience and consciousness. Today's more rule-bound ethics adjudicated by experts has pushed this more democratic approach out of official

198

public favour. Nonetheless, a goodly number of us seem to have an abiding faith in it. It's evident in the staunch support for whistle-blowers, in the public's continuing curiosity about the people who resisted Adolf Hitler's regime and, particularly, why they did so when so many others went along with the creeping totalitarianism or saw themselves as mere cogs in its bureaucratic wheels.

DIALOGUE AND THE ART OF ETHICAL DISCERNMENT

Being one of the curious myself, I reached for the books on my shelf by or about the German-born political theorist Hannah Arendt. The most recent, published in 2003, contains two lectures she wrote immediately after her book on SS Lieutenant Colonel Adolf Eichmann's trial for war crimes, in which she coined the phrase "the banality of evil" to capture what she had witnessed at the trial with her own engaged sensibilities and thought. Whereas others depicted this architect of the Holocaust as a monster, she saw him as ordinary, a typical bureaucrat who liked to follow procedure and keep careful accounts and for whom buzzwords and stock phrases substituted for thought of any depth.

In these follow-up essays, Arendt endeavoured to put her finger on why such evil as the Holocaust could happen and how it could be prevented. Again and again, she stressed the importance of self-knowledge and of an inner dialogue (between "me and myself," as she put it) in determining right from wrong, with a sense of inner harmony and living at peace with one's self following from having said or done the right thing. She stressed dialogue with others as a medium for nurturing this ethical or moral thinking. She drew on the example of Socrates who, in his insistence on taking philosophical discourse to the marketplace, also seemed to believe that everyone has the capacity for critical thinking and even, perhaps, the duty as a citizen to cultivate it. For Arendt, judging right from wrong doesn't follow from complying with some rule or abstract idea: it always depends on the context. That's why dialogue, particularly in the context of mentor-teachers and students, is so vital in public education. Arendt was a living example of this throughout her life, and so is Ursula Franklin, who has been my mentor and friend for more than twenty years, combining university scholarship with community engagement and sharing this with the likes of me through stories and probing questions in dialogue.

It was time for class. I put the book on the shelf. I gathered my notes and entered the classroom.

On the spur of the moment, I put my notes aside and simply asked my students: "Why are you here?" The discussion that ensued was straight out of our here and now.

The first student to speak is very clear: "Well," he says, "a lot of people think they're here 'to have fun,' but others are 'seriously preparing for a career.'"

Another student objects, saying that he rather likes the idea that the bachelor's degree he is pursuing is "less packageable." He argues that the critical-thinking skills he is acquiring might yet have a longer shelf life than, for example, a degree in information systems. "Critical thinking lasts longer," he says.

A third student states that she used to take everything she watched on CNN (TV's Cable News Network) at face value, but not any more. "University allows you to think past that."

Another agrees, "Every course I've taken has made me think."

In the end, the consensus is that they are there to think. Even the first student is nodding, seemingly having changed his mind.

We then talk about the changing university environment, including the productivity pressures on students.

"It's not about learning the course; it's about passing the course," one of them says.

They discuss the positive aspects of multimedia Web-based and on-line learning, how accessible this is, how much more choice it offers. But there are troubling facts. I tell them about the results from the academic-time survey, and how much more time this technology demands of a professor, not just in putting up and maintaining Web pages but in keeping up with e-mail from students around the clock. I share with them the findings of some research conducted by two colleagues at Carleton University, which indicated that students in fact didn't participate much in the on-line chat groups set up to support the broadcast lectures and/or Web-based learning modules. Even in courses in which a portion of the final mark was linked to participation in on-line discussions, the quality and depth of debate was poor, with very little sustained and developed argument from one posted contribution to the next.

Our time together is running out. The final comment comes from one of the women, who says that she remembers hearing the phrase "the personal is political."

"It's like they're taking the personal out of the political. They're detaching the personal in on-line learning." She looks at me, then shrugs as though it's just a passing thought. But as we're both gathering up our stuff I return her look, a look that acknowledges what she's said. A smile lights up her face, I suspect because I've just affirmed that her take on things matters, that it's worth listening to.

As I left the room, I felt that maybe I'd done my job for the day, because we'd just spent three hours putting the personal back into the picture. We'd spent three hours enhancing each other's presence in time. In a modestly furnished university seminar room, we'd cultivated a habitat of respectful, responsible intellectual engagement that cut across the biases of the larger environment outside, that resisted the fading out of these particular students' grounded identities, the desertification of the space of civic discourse.

Week after week during the course we'd built a memory of, and a comfort level with, communication in its good-faith sense of the word, and similarly we'd built the capacity and aptitude for dialogue and discussion. And all of this resisted the eclipsing of human agency in our society, resisted the habits of "disappearing" people from the here and now and replacing them with a virtual presence paced and governed by knowledge and information systems to the point that nothing much matters any more except speed and turnover. We'd resisted an attention-deficit culture being compounded into a democratic deficit.

There are thousands more like me, putting time back where it belongs, in the mouths and ears and yearning-to-be-implicated hearts of our next generation of citizens and community, corporate and political leaders. Perhaps there are tens or even hundreds of thousands of us, in schools, colleges and universities around the country. But we're tired, strung out and under attack. We need a break.

Nine

CIVIC DIALOGUE
AND NOISY SILENCE

"*Citizenship by proxy is an oxymoron.*"
ROBERT PUTNAM, Bowling Alone

"*If we organize our society so that we can't use
our common sense, then we become dysfunctional.*"
JOHN RALSTON SAUL, On Equilibrium

THE JOURNEY of this book took me to Walkerton, a town of 5,000 residents in rural southwestern Ontario that is known for its tainted-water tragedy. During the ten days following Mother's Day weekend in the year 2000, roughly half of Walkerton's residents fell ill from a virulent E. coli contamination of the town's drinking water. Seven died, 100 were left with kidney damage and 500 with chronic diarrhea.

When I arrived just over a year later, an inquiry chaired by Justice Dennis O'Connor was laying most of the blame for what happened on deep provincial spending cuts. Yet as I took my seat to hear the wrap-up testimony, I couldn't help wondering if there wasn't more to the story. Being able to deal effectively with society's moments of crisis and danger depends on being able to know what's really going on, what's worth paying attention to and what to trust as real. Equally, it depends on being able to come together in society, to engage as citizens and to act collectively in society's best interests. So I'd come to make some inquiries of my own. What did people in this town notice before or during the crisis? What actions did they take?

The inquiry had heard that under the so-called Common Sense Revolution launched by the province's neo-conservative Premier Mike Harris in 1995, not only had the government slashed 900 jobs from the Ministry of the Environment, including 42 per cent of the staff monitoring drinking water, it had also privatized the ministry's water-testing labs and eliminated a number of public-accountability measures such as ensuring that adverse test results were relayed to public-health officials. At the time, apparently, the minister of health wrote to the minister of the environment pleading that they fix the gap of communication that could leave public-health officials in the dark if something went wrong with people's drinking water. Nothing seemed to have come of this. (Testifying before the O'Connor inquiry, Premier Harris argued that a number of the concerns raised at the time these changes were implemented were never disclosed to the public because it was "information for decision makers." Technocratic decision makers, presumably, not the public or their elected representatives.)

Much of the inquiry had focussed on the actions of Stan Koebel, the general manager of the Walkerton Public Utilities Commission and, to a lesser extent, his brother Frank, who worked for him as a foreman; as a result of the government budget cutbacks, responsibility for the safety of Walkerton's water was essentially left on Stan Koebel's shoulders. Yet, as Justice O'Connor concluded in his report, the Koebels routinely "fictionalized" the data on which the town's drinking water depended. So how did these two brothers come to be the experts in charge? I wondered.

After high school, first Stan (who completed Grade 11) and then Frank (who finished Grade 12) went to work for the Walkerton Works Department, as the utilities commission was called at the time, where their father was a foreman. Both served linesmen apprenticeships for the electrical side of the utilities before moving on to the water side as jobs opened up. By the time of the tainted-water crisis, Stan had a Class 3 operator's licence for running the water system and the test sampling, though he had been "grandfathered" into this responsibility largely based on his work experience.

The inquiry had heard that part of Stan Koebel's job involved testing the water from three wells that provided Walkerton's water, to ensure that it was safe for drinking. Every day he was supposed to check the chlorine residual to verify that it was sufficient to neutralize any toxins in the water, and once a week he was expected to collect samples and send them away to a lab for more detailed tests.

For five straight days, from May 8 to May 12, 2000, Walkerton received excessively heavy rains (seventy millimetres fell on May 12 alone), which flooded the surrounding countryside and caused manure-contaminated water to seep into Well 5, the shallowest of Walkerton's public wells. Stan Koebel had left town on May 5 and planned to return on May 14, so Frank assumed his duties during that period. However, instead of testing the well's chlorine residual on Saturday, May 13, and again on Sunday, May 14, Frank simply entered in the logbook the number he usually wrote (.75 milligrams per litre), a number that indicated the water was fine. In his report Justice O'Connor wrote that "clearly" the residual entered "is false" and that the one entered for May 14 was similarly "fictitious."

When Stan Koebel returned to work on Monday morning he started drawing water from Well 7, although the new chlorinator for that well, which had arrived a year and a half earlier, had not yet been installed. (He later doctored the books to suggest that the chlorinator had been working and the water being duly chlorinated.) That afternoon, he also switched off Well 5. Meanwhile, the effect of pumping unchlorinated water was to dilute what little chlorine there had been in the system even further. Koebel was due to collect water samples from each of the wells in use and send them to the lab for testing that day but, as was his usual practice, he simply filled some of the bottles from the tap at the utilities commission office and labelled them as though he'd gone out to each of the designated sites. (It was a matter of "convenience," he told the inquiry.)

The following day, May 16, the lab called to say it had received an insufficient water sample. (As it turned out, since 1991 Koebel had repeatedly failed to send the amount of water the lab needed to complete all the required tests; a sampling audit had revealed this fact among other irregularities associated with the Walkerton water utilities and at the time Koebel had promised to follow correct procedures. However, an audit in 1995 found that he hadn't, and yet another audit in 1998 found the same thing.)

When results from the tests the lab could do became available on Wednesday morning, May 17, lab staff phoned Koebel to alert him to the news that three of the four water samples had tested positive for E. coli. They also faxed the test results that day. At this point, not only did Koebel fail to pass on news of the bad water to the public-health authorities, he lied to at least one person who phoned the utilities commission to ask about the water after noticing that people were starting to get sick. Koebel thereby misled the public into thinking nothing whatsoever was wrong with the water. He did, though, send his brother out to install the chlorinator at Well 7.

By Friday, May 19, people all across town were vomiting, and coming down with diarrhea and even bloody diarrhea. Those suffering included students in the local schools and the elderly in the town's two seniors' residences, Brucelea Haven and Maple Court Villa. That day, eight people with the same symptoms showed up at the local hospital and one doctor, Dr. Donald Gill, saw at his office at least a dozen patients exhibiting the same profile. By then, too, Dr. Kristen Hallett, a doctor in nearby Owen

Sound, had completed a food history of two patients with bloody diarrhea whom she'd seen on Thursday. Having ruled out food as the cause, she suspected E. coli O157:H7, which has been associated with raising livestock in factory farms and can be lethal. (It destroys red blood cells, including clot-producing platelets. It also damages the lining of blood vessels and can trigger kidney failure as the bloody debris clogs the filtering system in the kidneys.) She thought the drinking water was the likely source of the problem.

At nine o'clock on Friday morning Hallett phoned the local public-health unit to share her information with Dr. Murray Mc-Quigge, the medical officer of health. His executive assistant left a note on his desk summarizing Hallett's news and at some point also left a message for David Patterson, the assistant director of health protection at the public-health unit. When Patterson checked his voice mail at noon, he asked a public-health inspector in Owen Sound to follow up. By four o'clock Patterson had enough feedback, with specific finger-pointing at the quality of the Waterton water, that he called Stan Koebel.

Koebel had already received a call that day from James Schmidt, the public-health inspector in Walkerton. Schmidt was following up on calls he'd received from local citizens including Ruth Schnurr, an administrator at a school with twenty-five students off sick. Apparently Schnurr bluntly told Schmidt that she thought water was the problem.

When Koebel spoke with Patterson, he reportedly told him, as he had Schmidt, that "he thought the water was okay." So nothing official was done. Still, the townspeople did notice Koebel out with hoses, flushing water from the mains on Friday afternoon and continuing into the weekend.

When I spent an evening with some women who worked at Brucelea Haven and asked about that weekend, they recalled seeing what should have been an indicator that all was not well.

Marcie remembers: "We all saw it, the flushing. This is normal procedure if you've got a problem. And I can remember I was driving to work (late Friday afternoon), and the hose was coming out of the water main. We all witnessed it happening. We all saw it happening. But none of us twigged, not 'til later."

Marcie's colleague, Anne, who is sitting across the kitchen table, picks up the thread: "Someone was saying, 'If you ever see them doing that [flushing the water mains], you know you're in trouble.' But that was just chit-chat. People just talking, with nobody really knowing. Because nobody official had really confirmed anything.

"Our neighbour, who works with the OPP [Ontario Provincial Police], he was very sick, and he has lasting kidney damage from this; he tried to call the public-health unit on Saturday, and he got the voice mail. So he called the hospital and told them: 'It's got to be the water.' They listened to him, but they didn't act on it because they had been assured that it wasn't the water."

Marcie remembers that the residents at the nursing home had been coming down with terrible bloody diarrhea since at least Wednesday. Staff were falling sick too, or their children were. When Marcie got home on Saturday morning, her own daughter was sick, cramping so badly she passed out in the shower. Being a nurse, Marcie knew what medication to buy, but the pharmacy wouldn't sell her any of the regular anti-diarrhea medications, apparently at the recommendation of local doctors. The directive had been issued some time on Friday afternoon, it seems, yet the pharmacist either couldn't or wouldn't say why. (In cases involving E. coli O157:H7, such medications can exacerbate the effect of the toxic bacteria.)

"So," Marcie asks me, her voice rising, "if the doctors were saying that, why weren't they saying 'boil the water'?" (Why didn't she? I wonder.)

On Sunday, the local radio station was still saying the water was okay. However, by Monday, May 22, Marcie recalls, they could hear the helicopters taking some of the sickest people to hospitals in Toronto. Word spread that someone had died.

The Walkerton crisis is so frightening precisely because it was unfolding before people's eyes, ears and noses, in the bodies of their nearest and dearest, and simultaneously passing right over their heads. They had fragments of observation and experience, but the dots of information remained unconnected.

CIVIC ENGAGEMENT AND THE DEMOCRATIC DEFICIT

Walkerton is Canada's Chernobyl and the parallels run deep, including the nature of the threat to public health and welfare involved. Commenting on the experience of nuclear radiation at Chernobyl, sociologist Ulrich Beck wrote: "All of us—an entire culture—were blinded even as we saw. We experienced a world unchanged for our senses, behind which a hidden contamination and danger occurred that was closed to our view—indeed to our entire awareness." Like Chernobyl, Walkerton's tragedy involved a fairly new danger: E. coli O157:H7 is a mutation of E. coli bacteria that is considerably more dangerous than regular E. coli yet just as undetectable until it attacks the body. In such cases, perception of the risk has to be constructed for the public by technologies of detection and information-processing operated by experts in risk assessment and risk management—in short, by Harris's "decision makers." And that's precisely the problem, Beck argues. We, as "citizens, have lost sovereignty over our senses," and that, almost by default, can lead to a "legitimate totalitarianism" by the technocrats. As these experts goof up, burn out or betray our trust in them, we are left feeling helpless to help ourselves.

The bridge between the official data sets (the water's chlorine residuals in Walkerton's case) and reality is broken or weakened, often because in a global world they are so many times removed from each other. So, when something terrible is happening and the knowledge systems fail, we can lose our ability to seize the initiative and act in our collective self-interest because we cannot rely solely on our personal experience. Unless there's a compensating dialogue between the experts, various social institutions and

209

the general public in which the data can become real and public trust reaffirmed, we risk apathy or hysteria and what some cultural critics describe as a pseudo form of schizophrenia: a derangement in our powers of perception and communication, of our powers to articulate what is real. When we can no longer distinguish between delusion and reality, we can't trust anyone, not even ourselves. That's part of what happened in Walkerton.

One reason that people don't seem to act is because of a withering and atrophying of their ability to bring all the data, all the complexity, together, interpreting it in real-life contexts so that they can take innovative and appropriate action. As political philosopher Thomas Homer-Dixon states in his award-winning book *The Ingenuity Gap*, there is a "very real chasm that sometimes looms between our ever more difficult problems and our lagging ability to solve them." For me, it's a case of too much data, not enough dialogue and also, perhaps, what German linguist and George Orwell scholar Uwe Pörksen calls "plastic words."

The key to plastic words, such as "developments," "norms" and "indicators" is that they often come from the world of science, technology, economics or administration, and as a result they carry with them an aura of authority, rigour, objectivity and unassailability. The trouble, Pörksen writes, is that they're then transposed into everyday speech by public officials who use them as a stand-in for detailed explanations. Because such words have no intrinsic meaning, they can be used—by varying their tone—to impart an instant understanding that everything is under control and double-plus good or, on the contrary, suspect and double-plus bad. Like statistics, these words are part of a new global language of public-issue management that is "unhinged" from a historical context. Like brands and logos, they are a language of ready-made meanings that can be mobilized in an almost virtual form of public dialogue while masking the fact that very little has been communicated, explained or justified. One reason politicians (and their technocratic assistants) use this language is because they are so stretched for time, attending as they do a succession of short meetings, briefings and photo opportunities. And one reason they can get away with glib, mix-and-match sound bites is that the public isn't involved enough in debates of the day to notice or even much care.

This is linked to what has been called the democratic deficit. More urgently still, I think it speaks to the crisis of meaning in society and the reported withering of public trust and growing sense of disengagement. If we are to invest ourselves in shared meanings, and thereby take meaningful action as a society, it matters how we live together. To build trust, to challenge and check out the truth of what we are told, it matters how we are engaged with each other. Taking responsible action requires direct participation and an educational and even cultural tradition that encourages and champions this broader dialogue, plus, of course, the time and circumstances to practise it. In short, it requires a context of civic engagement—that is, of involvement in the common welfare of society. It is citizenship pursued mostly outside formal politics, in community and non-governmental organizations but also in more informal and impromptu discussions among co-workers and friends. It is particularly relevant in public institutions like schools, hospitals and doctors' offices that are rooted in the community, because it is about being an active subject in the business of one's society, not just in the events that become headlines. Like any other social skill, this level of involvement has to be practised regularly to remain an integral part of a person's sense of self and how the world works. But as I learned when I dug deeper into the subject, it is not.

Robert Putnam, a professor of public policy at Harvard University, has found that, with the dramatic exception of evangelical Christian churches, most forms of civic engagement have declined over the past fifty years. In *Bowling Alone* he draws a link between healthy societies and societies rich in social networks and civic participation, or "social capital." He also distinguishes between two forms of engagement: "Social bonding" can bind people, often rigidly, into a homogeneous group identity and mission, whereas "social bridging" allows people to come together as a collectivity, negotiating consensus and compromise while preserving their differences. Championing the latter over the former, especially in our modern, globalizing world, Putnam defines it rather delightfully as a lubricant, a sort of "sociological WD-40." Strong institutions of public culture such as public schools and universities are important ingredients, but so too are an independent civil service, labour law and other measures to promote justice in everyday life. Given opportunities for meaningful participation, the literacy people gain at school and university can mature into civic literacy and the ability to speak ethically with "a civil tongue."

In a book by this title, which extols the virtues of ethical citizenship through public dialogue, philosopher Mark Kingwell draws on such seemingly old-fashioned virtues as good manners to emphasize the disposition to others and their differences that is necessary to work out values like justice and fairness in context. Civility, or treating the other person "as if they were worthy of respect and understanding," is essential, he says. In Kingwell's view, civic dialogue is public conversation governed by an ethic of caring, tact and even, if possible, "the rare gift of empathy." Potential conflicts that arise from differences of perspective can be leavened by the daily bread of politeness and tact, that is, "a trained receptivity to the otherness" of other people, ideas or beliefs. Moreover, he believes that a consensus on what matters to us as a society can and will emerge from this down-to-earth talking about what matters to each other.

THREATS TO CIVIC ENGAGEMENT AND DEMOCRACY

These days it's easy to be a technocrat, a fast and agile mind immersed in the hypermedia zone, deploying technical expertise, rules and data as, for example, "health indicators" or the "latest developments." It takes effort to remain a whole human being immersed in society, thinking and speaking for ourselves as citizens and with a duty, both to ourselves and to society, to do so. It takes time and commitment to the larger picture to keep restoring "developments" to the social and historical context, and to ensure that the health indicators being adopted reflect the public-good goals of a healthy society. Yet the routine opportunities to think and talk, the shared time and space and the pace required for people to listen politely to each other, attuned and possibly empathetic as well, are going missing these days. They're disappearing not just in people's workaday lives but also in what is loosely considered the public sphere, where we trade notes on the larger wind shifts of change and what they portend, where we decide what should be done and/or agree to be governed by the consensus.

Putnam uncovered several reasons to explain the drop in civic engagement. For many, the stressors of being too busy, always rushed and working "very hard most of the time," preclude participation. Long commutes to work (seventy-two minutes a day on average), usually alone behind the wheel of a car, limit others. Putnam also identified generation-specific trends, such as the fact that post-war baby boomers and Generation X watch more TV and use computers more often, which disallow direct involvement, whereas the generation that came of age during the Depression and World War II has experienced not only shared public events but has seen its personal involvement make a difference. Furthermore, he reports, smaller communities and social institutions are "better from a social capital point of view" in that they make civic participation more accessible and real. He notes that when locally owned businesses and locally run public and private services close their doors some traditional forms of civic leadership disappear, along with reasons for coming together in local communities.

Putnam also speculates about a new virtual, pseudo form of civic engagement, a sort of "proxy" participation through direct mail. He notes that people who join organizations that use this strategy don't have to attend meetings, contribute to the dialogue or participate directly in actions arising from it. Responding to direct-mail and telemarketing solicitations, they send in money for memberships or special campaigns and receive newsletters that report discussions and actions taken by paid staff and some project-specific volunteers. The result, writes Putnam, is a sort of "symbolic affiliation." That is, their relationship to the organization is more simulated. Their participation is also more volatile and fleeting; he cites the low membership-renewal rates in many organizations, particularly after a high-profile event has concluded or an image-damaging story hits the news, to illustrate this point.

Urban sociologist Barry Wellman, of the University of Toronto, has studied changes in people's involvement in networks and communities in Canada and found similar patterns. He has traced the shift from group participation in shared time and space to selective, often short-term participation in networks. These are often functional networks, he notes, in which people come together, often virtually rather than directly, in specialized roles as "fragments of selves" (for instance, as user groups, consultants on contract, video-game players or parents of children with asthma), with "multiple, limited commitments" to many, narrowly focussed tasks, rather than to the big picture. In these networks, too, place is less important than the flows of communication through space, which is itself more and more contextless. In fact, Wellman writes, many geographic spaces have become merely places to pass through or touch down, to pick up groceries or dry cleaning. He describes direct communication as similarly "residual" in shared time and space, and he suggests that while people still live in geographic communities and have ties within them, "these ties are only weakly connected."

My own belief is that although on-line networking and face-to-face communities are often depicted as opposites, they work best as a blend. As Putnam says, "An extensive, deep, robust social infrastructure of relationships must exist so that those using the electronic media will truly understand what others are communicating to them." That is, through personal reflection and sustained dialogue with others both on-line and off, people can make the

214

ephemera of mere data real and hold the borderline fictional to account as they interpret it in a context of shared understandings and mutually trusted words. But it requires regular conversations to temper the simulated and abstract world in which, without even realizing it, people can become dissociated from the consequences of a phone call left unreturned or a data field left incomplete. Dialogue remains the foundation of civic engagement; it both helps people to feel fully present and to take responsibility, and challenges them to do so.

LOCAL DIALOGUE, LOCAL CONTEXT, LOCAL ENGAGEMENT

Before leaving Walkerton, I visited one more person. Lisa Stroeder is the twenty-seven-year-old clinical nurse at the Brucelea seniors' residence who assumed some responsibility over the drinking water, and acted. Late on the afternoon of Friday, May 19, while the fax confirming the truth about the town's toxic water sat somewhere at the local public utilities office, Lisa switched the residents from local tap water to bottled water. I wanted to know why. I drove out to the farm where Lisa lives with her husband and two small children and I simply asked her to tell me what had happened.

She begins by explaining that it had rained and rained throughout the preceding weekend, causing extensive local flooding, including near one of the town's wells, where some road construction was going on at the time. A day or two later, her husband came home reporting that his boss's daughter was sick with bloody diarrhea. This triggered in Lisa the memory of when her own daughter had fallen ill less than two months previously. She'd been going to a babysitter in Walkerton at the time, and what Lisa remembered was the doctor specifically asking if the diarrhea was bloody and suggesting that they test the farm's well.

By Wednesday of that week, the seniors at Brucelea were sick with the same symptoms. Following the standard procedure for whenever there seems to be a pattern in illness, the nurses started line-listing all the patients with common symptoms.

215

At two o'clock on the Friday afternoon, Lisa was doing this with one of the other nurses. The nurse told Lisa about a family whose child had been sent to a paediatrician in Owen Sound, and she said the paediatrician was asking hard questions about the local water. The young mother had spoken to her own mother over the phone about this situation, and she in turn had mentioned it when visiting her mother at Brucelea. The great-grandmother had passed on the news to her nurse, who mentioned it in her chat with Lisa. That's when everything started to jell in Lisa's mind.

She recalls: "So when N. told me that, around 2:30 or something, and I knew the family where this girl is sick and you've got everybody on staff saying: 'My wife is terribly sick... my next-door neighbour is so sick she went into Emerg last night.'

"It's a big employment here in Walkerton, Brucelea Haven. There's a couple of hundred on staff. So in a small town with all your neighbours and people with small kids and everything—and that was another thing. We knew that a lot of kids were going home from school sick."

She continues, "And I'm going, 'Oh-oh, it's Friday. It's nearly three o'clock; that means public health is shutting, doctors' offices are shutting. I'm going to Toronto for the weekend.' So I'm going, 'Oh-oh, if we've really got a problem here, I've gotta act fast on this.'

"First I phoned Dr. Gill. He's medical director (for the residence). And I told him, 'We have some bloody diarrhea: three on one floor.'

"And he said, 'That's all I've had at my office all day. Different age groups.' And I asked him, 'Did you hear about the school?' And I told him about the girl in Owen Sound.

"He said that public health had contacted him. They were calling all the doctors, and they were supposed to be on the lookout.... They were supposed to keep track of the cases and report back, sort of thing.

"So I asked him, 'Should I be boiling the water?' and he said it probably wouldn't be a bad idea. Because before you go and do something that big, you don't want to seem off-the-wall or whatever. So then I got off the phone with him, and it was getting around four, and public health in Owen Sound shuts down. So I got on the phone to someone in Owen Sound… and finally I did speak to a nurse…. I asked if we should boil the water, and she put me on hold a few minutes, and she said it might not be bad as a precaution.

"And afterward we were wondering, well, if we got told to do this as a precaution, why didn't everybody? And I think she was really stepping out to say that. Because she was only gone [from the phone] two minutes, and [Dr. Murray McQuigge] might have already been gone. [The O'Connor report is silent on any actions taken by McQuigge, the medical officer of health who was officially in charge of such matters, assuming that he was around that day.] You know, it was the end of the day before a long weekend."

By then, Lisa recalls, it was 4:15 p.m. Across the hall from her office, she could see the water and juice jugs set out on the supper tables, ready for the first sitting at 4:30. She sat there staring at them, and that did it.

She says: "I just went down to the kitchen and told them to get the stuff off the table." And everybody did. They also went from floor to floor, dumping all the water jugs and even the ice-cube trays and distributing bottled water. (They missed one ice-cube tray; it was enough to make one elderly gentleman very sick.) Some even stayed late to boil water for cooking and cleaning that night and for the following morning.

I was curious as to why people pitched in, when Lisa lacked the authority to tell them what to do. She thinks it's because she was simply stating the obvious. As she put it: "Everybody in the kitchen knew people who were sick. It was very real. So it made sense to start boiling the water."

It was "common sense" in the original meaning the essayist Thomas Paine used in 1776 to articulate the consensus of American civil society: that it was time to declare independence. Just as

217

Paine "embodied that [sense] in the clear, forceful language of the people," I think it was Lisa's being able to both connect the dots of evidence and communicate what they signified in the language people use every day that prompted them to act.

I take her back over exactly what motivated her to act, and we tick off the factors:

· the context of a close community where everyone knows everybody else, and stories percolate as shared experience;
· the formal and informal dialogue between herself and the other nurses;
· the period of time over which things started to add up as a reality clear enough to recognize.

It resonates with what urban planning professor Jane Jacobs has written about the importance of shared time and intersecting paths of social circulation in the economy of cities. For Lisa, too, a key was having enough discretionary time that she could stop and really think.

"Normally you just run, run, run," she says. "To really stop and think about the situation, that's one thing to my advantage in having an office. I can sit in there and say, 'Okay, Lisa, what are you going to do?' Whereas the nurses on the floor, they have the constant chatter and the noise and the people and the tasks to do.

"It's a good thing too about my naïveté about the water system. I didn't think, 'It can't happen if it's all being chlorinated.' I just thought, 'Well, anything can mess up, especially after the flooding we had.' That was going through town. People in town were saying the water was contaminated.... "

Suddenly Lisa's crying, motioning for me to turn off the tape recorder. It seems that some people in town turned against her. They thought that maybe she had gained access to the official data confirming the E. coli contamination and she hadn't shared this information with the larger community. Perhaps they couldn't believe that one person could act on the evidence of their senses and the shared knowledge base at hand, when they themselves could not.

(Apparently, too, Stan Koebel didn't understand the significance of the knowledge that was lying on his desk. As his lawyer argued in defending the Walkerton Public Utilities Commission general manager, he didn't really understand the biochemistry involved in clean drinking water. If it still looked okay and smelled okay and tasted okay, that was good enough for him.)

So the tragedy of Walkerton takes on the aura of a symptom, suggesting a societal dissociation, a collective disconnection from lived reality. And it foreshadows the consequences if we lose our ability to act on what really matters in life. Says Robert Putnam, "Disengagement and disempowerment are two sides of the same coin."

LESSONS FROM SILENT SPRING

In 1962 American biologist Rachel Carson published her classic wake-up call, *Silent Spring*, a book that brought the dangers of DDT and other pesticides to life by describing how they were poisoning frogs and birds and other creatures and threatening them with extinction. Writing as a citizen, not just as a research scientist, Carson had decompressed the data, breathing life and feeling into the facts and interpreting them in the context of nature itself. Unless measures were taken to ban and restrict the widespread use of these lethal chemicals, Carson predicted, whole species of birds that usually sing us into spring would die.

Now, I sense a variation on this fate. We could be losing the power that her book, and its success, proved we still have: the power to wake up to urgent realities around us, and to act. The birds might still be singing, or singing strangely, but we might no longer really hear them because we're not present enough or paying

219

attention enough to our senses. We're too stressed and fatigued to notice anything but the immediate task in front of us. If we're too preoccupied with the official indicators of life and death to take in what our bodies sense, including the birds outside, we might no longer respond to messages, or even the song of life itself, at that level. We might no longer take our cues on what matters from the lived here and now, or use it to check out the data.

Chapter by chapter, I have elaborated on how we could be losing this feeling for ourselves, and life: not just as compressed time and space delivers us into an experience of life as fleeting, fragmented, superficial and subject to change at any moment, but equally as an atrophying ability to credit what we see and sense going on around us, even in our own bodies, and to interpret these and other signs for ourselves. We could be losing the capacity to feel engaged and implicated enough in the larger whole of society to act on a shared sense of the public good, to seize some measure of public initiative, to save the birds, to save the children, to save whole generations of people in Africa dying of AIDS, to heal the injustice of deepening inequalities locally and globally, to intervene when people are getting sick and dying and it looks as if the local water is contaminated.

Stan Koebel and his brother Frank are hardly in the same league as the consultants associated with Enron's virtual-empire building and related "black box" accounting, or the journalists whose fraudulent work has appeared in the world's most prestigious newspapers. Perhaps that's why this instance of data doctoring warrants special attention: precisely because it is so ordinary and everyday. When I set out to write this book, I was surprised at how often the subject surfaced, by the by, in conversations: truckers cheating on their logbooks; teachers, social workers and health-care workers playing the categories on the performance records some even describe as "cheat sheets." Data massaging and spin-doctoring have settled into the social and cultural environ-

ment. The important point for me is not just that this is dangerous, but that it's almost inevitable as people become anaesthetized
by the pace of life and by the symbols that, travelling at lightning
speed, propel it.

The greater danger, then, is disengagement. We are allowing
the technocrats to take over public debate and policy-making, as
fewer and fewer of us turn out to vote or challenge the Commons
committee or the staged townhall-style meeting to be meaningful
components of democratic decision-making. As we feel more and
more removed from the reality of the statistics we are fed, we are
becoming increasingly apathetic and that's why so much data can
be fictionalized.

I don't want to read more into Walkerton than there is, nor
treat its citizens' tragedy as a symbol. (In fact, that would betray
one of the lessons I have learned from writing this book: the danger of letting symbols supply ready-made chains of significance,
instead of interpreting things for ourselves, smithing our own
words and symbols out of our own forge of time, in dialogue with
others in the institutions where we both work and are involved in
various communities.) Nonetheless, the events in Walkerton in
the spring of 2000 suggest to me a variation on Carson's prediction in *Silent Spring*. Poison in the drinking water literally silenced
at least seven lives and crippled many others, but more telling is
that people did little or nothing to help themselves and address
their shared problem collectively. The silence in Walkerton wasn't
a real silence, of course. It was a busy silence as people stumbled
around, sick or tending to the sick; swapping stories on the phone
and on the street because they could get nothing but recorded
messages from the official decision makers (none of whom, it
seemed, lived in the town), sharing information and trying to
make sense of it all yet never organizing a public meeting to dis-   221
cuss this knowledge and take collective action. It was a noisy silence. The noisy, frantic silence of a dysfunctional society.

Part FOUR

RENEWAL:

Reclaiming a Feeling for Ourselves

Ten

TAKE YOUR TIME

"Balance and harmony are experienced, not seen."
VANDANA SHIVA, Staying Alive

"There is a time and a pace for everything."
BARBARA ADAM, CBC Ideas

As I REACH to get the bowl from its resting place on the cupboard shelf, at some point in the bowl's descent voluntary simplicity suddenly makes sense. The bowl sits comfortably in the spread of my hand as I carry it to the counter, as I've carried it for well over twenty years, carefully packing and unpacking it through at least three moves and all the ups and downs of married life, including the final move when my marriage was over. It's a medium-sized pottery bowl, so perfectly rounded that if I had two and stuck them together they'd happily roll across the floor. Turning it over to check the artist's name, I see the spiral marks from the wheel that spun this shape into being under the wet and steady hands of a woman in Dundas, Ontario. She left a lot of the natural clay colour to glow through the glaze and decorated the side with simple motifs: what look like a series of butterflies and a stuttering border of luminous lapis lazuli. When I get out the bowl, I often think of my dear cousins Jennifer and Jessica Bayne, who gave it to me as a wedding present. I also remember how I used it once to make cherry Jell-O when my son Donald had his tonsils out, and how many times I've served beans or broccoli in it when friends have been over for dinner. As I hold and heft its familiar shape and weight, I realize how much it holds me, too. With the memories spilling over the blue-limned top, I realize how much it plays me back to myself. As it does, it also slows me down as these associations catch my attention.

I'd always thought the voluntary simplicity movement was just about rich people jettisoning excess stuff in life and feeling smug. Still, I could see the merit in paring down the things you have to buy and throw out, buy, upgrade, maintain and replace, plus worry about having stolen in a break-in. I've become a convert now that I realize that voluntary simplicity is also about recovering a sense of self in local and personal things, and feeling anchored in place and time through them. We end up wanting less because we get that much more out of what we have. What philosopher and literary critic Walter Benjamin called the "aura," the "authentic" presence of an object imbued with the character of a place because the people who made and used it wove their particular talents and traditions into it, applies also to people. It's through time, then, that objects and people take on the distinctive character of place. Of course, this is precisely what gets lost when objects become mere commodities in the standardized space-time compressions of the global economy. And it's our individual selves that get worn away when we're going too fast, on-line and off, buying into new identities and investments, getting more things, getting more things done faster, processing our lives in a blur of fast-forwarding efficiency and servicing all the technology.

As I peer inside the bowl, I see a slight abrasion in the glaze at the bottom. I've left my mark, I realize, the mark of my serving spoon and my mother's Depression-era thriftiness in scraping the bowl clean every time. In a way it's like looking in a mirror.

226

ZEN AND THE ART OF SLOWING DOWN

Not long ago I read an article about the Japanese tea ceremony, or *sadou*, and how it can help to overcome stress. It's a lovely essay, describing and explaining the philosophical precepts behind every step in the traditional ceremony: the gathering of the few familiar things for making the tea, the importance of soft lighting and delicate incense in a place set aside specifically for this purpose, ideally with a low doorway so that everyone bows his or her head and enters humbly, as equals. Inside the room, the meditative anticipation while waiting for the water to come to the boil and for the tea to steep is just as important. The idea in repeating the same simple motions each time is not to perfect the technique and the task but to go beyond the details, to go with the flow of tea time and re-embed one's self in the primal rhythms of life. The ritual serves as a medium for slowing down, a meditative process for transforming clock time into lived time.

By participating in the tea ceremony we reweave space and time according to our own felt rhythms, such as breathing. We reconstitute ourselves through the rhythms of the ritual involved and emerge with our minds "cleared," our awareness "sharpened" and possibly, too, with renewed energy to face the responsibilities of our lives. Each time we do it we are also reminded that we can live consciously, conscious of ourselves and the values we live by, fully present and attuned to life around us. We are agents. We do have a choice.

I have to confess that I haven't quite succeeded in creating a proper tea ceremony, as I sometimes lapse into multi-tasking while waiting for the water to boil and the tea to steep for a whole four minutes! Still, if there's a zen, a meditative aspect, of tea, we can bring this mindset to other parts of our lives: to walking and swimming, to taking a bath, to hanging the laundry outside, particularly on a balmy spring day or even on a glittering-cold January day when the sheets may instantly stiffen like cardboard, then gradually soften as the ice crystals miraculously evaporate, leaving a residue of fresh air. Meditation can take many forms. Some people sit in absolute quiet and focus inward, "becoming intimate with your mind," in the words of one practitioner, becoming "less wired" and more "in touch with myself," in the words of another. Others use music as a medium, either making it themselves or listening to it (sometimes even to melodies specifically selected or synthesized as "sonic therapy"), attuning (even entraining) the vibrational rhythms of their bodies to those of the music.

We can slow down through simple, everyday things such as chopping vegetables by hand or making bread the old-fashioned way, with one's hands kneading the dough back and forth, back and forth on a flour-dusted kitchen counter. We can reconnect with ourselves by spending time with our children when they are young. I remember swimming at a public pool one winter's day and watching a man and his son sitting on the side of the wading pool. Again and again the man bent over to fill a green-and-yellow plastic watering can, then he poured the water onto his thigh just in front of his child. I watched enthralled each time the infant's hands came up and played with the falling water as if it were the strings of a harp. Once the water was gone, he looked at the drops still lying on the skin of his father's bare thigh. He patted the water with the flat of his hand, sending beads flying in all directions. The water all gone, he looked up, and the father bent to fill the can once more. As I watched this simple drama, I imagined the action taking the father back to that basic space-time, and I realized that my own rhythm was being slowed down too as I swam my lengths.

Now that my own child has grown up, I find that going for long, solitary walks works well (bicycling or rollerblading are okay, but walking still is the simplest). If possible, I choose a route beside a river, a lake or even an ocean if I can be so lucky, where the wash of water pulls me into the rhythms of tidal time and of my own beating heart. I find that I need at least an hour every day because for the first half hour I'm still processing all the stuff I

228

couldn't quite leave behind when I turned off the computer and left the house. I will briefly notice, say, a bit of acid-yellow lichen on the side of an elm tree, but so often I'm seeing only like a camera: merely processing information, not really taking it all in. But if I keep on walking, if I consciously focus on my breathing, almost using its rhythm as a mantra, I can feel myself drop back into my body, feel myself come awake inside. My eyes begin to "feel" what they see. It's almost as though my whole being reaches out through my eyes when I'm into the slowed-down phase of my walk. I don't just see, I understand, and I feel for each thing that I'm looking at. I am present, fully present to myself and to the universe just beyond my skin.

A VOICE OF ONE'S OWN

For Canadian poet-philosopher Dennis Lee, reconnecting with himself is also a way of tuning in to the deeper rhythms of society. He came to this realization as he grappled with trying to articulate an authentically "Canadian" voice from within the felt experience of being totally colonized by American culture and its ubiquitous commercial productions.

For Lee, the available language was making him feel choked and cut off from a sort of deep-time matrix and, as a result, he found it impossible to speak genuinely. His only option, he felt, was to drop out, to fall silent, withholding compliance and complicity. And in that silence, which was at least a refusal to participate in the prevailing commercial chatter, he began to hear what Canadian philosopher George Grant called "intimations of deprival." In other words, intimations of what was being muted and marginalized. Lee's writing now comes from what he calls "cadence," a particular rhythm associated with a half-buried life force, and what he calls "my body's mind," that is, his ideas coming from his body, not separate from it. His poems fiercely critique today's virtual consciousness of brands, statistical indicators and plastic words.

229

As moved as I am by Lee's messages, I am more intrigued by this poet's choice of medium, because it reminds me of the consciousness-raising circles I was part of in the early stages of feminism, and how we challenged patriarchal assumptions by telling stories based on paying attention to the often silenced voice of our bodies. I sought Lee out, and we met for brunch in his favourite local diner in an old and not-too-gentrified part of Toronto. The atmosphere was noisy with canned music, not conducive to quiet conversation, let alone recording one on tape. Still, we settled in, conjuring something over the runny eggs and steaming coffee. I leaned in close, and we talked about tuning in to the body.

"When I'm able to find time—deep time, not just chronological time—I try to let myself be caught up by cadence.... The naked existence of what is, that's what you tune into," he tells me. "It's a deep rhythm that speaks to the tap root of my being." To reach that point requires reserving a lot of discretionary time for himself, he says, and usually also involves certain personal rituals for slowing down.

"Such as?" I ask, and he explains that he listens to music, an eclectic mix of African drumming to classical to Charlie Parker blowing the saxophone. Then he can bear witness to "the jangle," his word for the chaos of everyday life in the larger world. "Ideally, you can report on that jangle if you're well attuned," he tells me. His way of doing this through poetry is by enacting the "body's sense."

He adds, "Does that return us to rhythms appropriate to paying attention in the rest of our lives? Will it resonate in the rest of our lives? I don't know. One can hope, that's all."

It's my hope as well. It's how I live my life as much as possible these days, notwithstanding my ongoing struggle to make ends meet, and to be present with my son and others close to my heart. The medium is the message, I've learned, and the truth of this applies both personally and politically. It matters how we live in space and time. The scale, pace and patterns (complex, constantly changing or simple and enduring) of our existence fundamentally shape us, affecting how fully we are present and what we consider to be real.

DECOMPRESSING TIME: A CONVERSION PROJECT

Decompressing time is perhaps the hardest thing to do. We know that if we live too fast, life can be virtually wasted; still, it's hard to find the inner throttle and, having found it, to voluntarily take one's self out of the fast lane. For me it's simply a daily struggle, and a choice. Every day, I go for that long walk. I usually don't work past five o'clock at night or on the weekends, and if I do I pay dearly. I've regained enough full-bodied consciousness to know that if I lose my inner serenity, my capacity to experience the full amplitude of the moment, I cheat myself and my work in ways that cannot be measured in money. Some people are involuntarily slowed down, by a heart attack or an illness that they sometimes end up honouring as a "wake-up call." Through their brush with mortality they realize this is all there is: life lived through the body in full awareness. They've gained a second chance to learn to slow down.

"Slow" has become the latest in cool these days, with a "slow food" movement championing the engaged sensuousness of meals made by hand, and a "slow cities" movement consciously preserving a historically slow pace associated with craftsmanship, religious orders or aboriginal communities still living on the land. Describing Haida Gwaii off the coast of British Columbia, poet and novelist Susan Musgrave wrote in the *Globe and Mail*'s travel section, "They say that truth moves to the heart as slowly as a glacier—that's how time moves here, also."

I suspect, however, that the most effective slowing down will happen as people change the micro-environments of their daily lives and, as *Globe and Mail* columnist David McFarlane suggested, "dare to do one thing at a time." In my own life, I think of it in terms of a conversion project, converting little bits of fast time into slower ones. I tell myself, for example:

· Take longer bathroom breaks.
· Actually take a break while having a "coffee break."
· Don't work through lunch.
· Stare down the inner taskmaster that says if you're not multi-tasking you're "wasting" your time.
· Be still, go for a walk instead of going to the gym, and walk at a leisurely pace.
· Once a day, if possible, choose one simple chore-like thing to do—weed the garden, iron clothes or make soup from scratch.
· Once a week, have lunch with a colleague or friend; or make a ritual of making tea or preparing supper together, and talk about this.
· Take time to be fully in touch with yourself when you're talking.
· Let yourself be vulnerable, uncertain, human.
· Make time for the little things, including the grace notes of friendship, such as phoning to say hello.
· Designate one day a week as a day of rest, with all or most gadgets switched off.

It works. Conversion can happen, though not by treating these as a to-do list. I have to let the simplicity and slower cadences of these time outs draw me in, converting externally paced deadline time into a life tuned to the pulse of my body and of my relationships with those around me. It requires constant vigilance and focus—the very things that are most vulnerable in our fast-forwarding, attention-grabbing, hypermediated social environment. But it's worth it. As these actions have become a regular part of my life, I've rediscovered the joys of discretionary time, the freedom of spontaneous chats and games, the pleasure of just being with my son or my friends and going with the flow.

232

Reviving the dream of a leisure society means reviving a leisurely pace, in one site of our daily life after another. As the micro-environment we create around ourselves settles down, as we let our senses dwell in the moment and feel our consciousness gain depth, we build a collective taste for it and the will to struggle for it, even build it in the places where we live and work.

SOCIAL MOVEMENTS AND RECOVERING THE TONGUE OF TIME

I expect the will to change things as a society will emerge only when enough people take the time to stop and really pay attention to their experience and the stories of the people they know and care about. It will begin when a critical mass of people realizes that time does matter—not just how we "spend" clock time, but how we live in time. Clock time is merely an artifice, a device used to organize, schedule and regulate life. Seconds and nanoseconds are mere abstractions. Taking our time means articulating the embodied experience of life. It doesn't just mean designating time *for* dialogue. It means living the time *of* dialogue and making that face-to-face experience real, which includes respectful listening, searching for commonalities, reconciling differences of opinion and moving toward possible consensus.

One reason that the business model has gained such an upper hand in the running of society, not just the economy, is because the language that helps to keep the social context in the picture, the language of experience and of talking about experience, has been pushed to the margins. Not only have statistical indicators replaced stories for adjudicating accountability, the language of systems and market economics has eclipsed the arts and the humanities, taking a sense of the lived and living context with it. Moreover, given the pace of life that globalism has inculcated so far and wide, there is no time for dialogue and little discretionary time and space for it, either. It's beside the point. So people keep silent even about the stress they are under, the broken sleep, the short-term memory loss and the borderline depression from never

233

catching up. In a sense, they assist in their own disappearance as human beings.

The social-protest movements of the 1960s reclaimed time as personal experience and story, and many of the actions that emerged from that time, including the "teach-ins" and "sit-ins," are being revived today. Some people follow Dennis Lee's and the hippie generation's lead, refusing complicity and compliance by dropping out. Many join the anti-globalization movement, living in sometimes shabby shared housing, or in tents or camped out in gyms or church basements. For example, those who went to Clayoquot Sound in British Columbia in 1993 didn't just live *at* the peace camp established to block clear-cutting in one of the world's last temperate rain forests, they lived the peace camp. The campers slowed down time in their daily activities—making bread and food together, learning in workshops, speaking in talking circles and staging various actions—and they became more attuned to the time of the forest around them: an ancient and even sacred time of interwoven, interweaving cycles.

In Brazil, some of the first initiatives in participatory democracy are starting to move off the street and become institutionalized; for example, in a participatory budget-making process adopted in some cities. The World Social Forum, a social counterpoint to the World Economic Forum at the Waldorf-Astoria Hotel in New York City, was convened in Port Alegre, Brazil, in 2002, drawing thousands of young activists from around the world to talk about genetically modified food, commodified water, community health and midwifery, spirituality and values. Writing about this event in the *Globe and Mail*, Canadian activist-author Naomi Klein described how an "indie media" member used her cellphone to relay the latest news from the Waldorf-Astoria to a micro radio station at the social-forum camp, where it was translated into Portuguese and broadcast at the forum and through local micro radio stations in Brazil. Klein ended her column by saying that "the heart of the World Social Forum was in unscripted moments like when my Italian friend Luca Casarini tried to sum up the summit over dinner: 'It's about—how do you say it in English? This.'" He was tugging at his T-shirt, pointing to the seam connecting front to back. "It's about the seams, the in-between spaces with their hidden strength," Klein concluded.

It's equally about time. It's about stitching time and space together differently and realizing that you can because, as the young Italian realized, we are the seams, and we stitch ourselves together through shared experience. As we engage with others in speaking our minds on the issues of the day, locally and globally, discussing what action should be taken, we constitute the bridges of social bridging, the bonds of social bonding and solidarity.

I felt the power of that hidden strength in Quebec City in 2001, where I attended the People's Summit and various street events organized as a counterpoint to the Summit of the Americas, a gathering of hemispheric leaders to discuss a possible free-trade agreement for the Americas.

On the Saturday, I tagged along with some female friends to the People's event at which Jose Bove, the farmer who destroyed genetically modified maize crops in France, spoke prior to a huge planned march, as did activist and Council of Canadians chair Maude Barlow and others. When word travelled down the line that the march was going nowhere near the chain-link fence that cut off the public from where the world leaders were meeting, I joined social activist and spiritual author Starhawk and a few dozen others in a spontaneous breakaway event. We had to go to the fence: It represented the brutality of the new world order being imposed in the name of free trade. It symbolized the lack of dialogue and negotiation.

Sharing water and vinegar for soaking bandanas and scarves, we headed up the steep hill toward the fence, where helmeted, gas-masked and well-armed police reinforcements guarded the many spots where protestors ("rioters" in most media reports) had broken through. Standing shoulder to shoulder, the police fired tear-gas canisters at anyone who dared to get too close.

We approached the fence, singing songs from Friday's joyous and life-affirming "Rivers of Life" event, which had happily indulged my taste for drama while making a plea for the world's fresh water. As each new wave of tear gas came at us, we ducked into house-lined culs-de-sac (where many residents had put out water hoses and buckets so that protestors could flush their burning eyes) to talk about what we should do next. Still we kept moving forward until, eventually, we found ourselves within sight of a spot where some thirty police officers plus a front-end loader were guarding a breach in the fence. Still singing, we advanced. As we did so, we heard a voice warn over a megaphone that we should turn around or we'd be shot at. We stopped.

Someone suggested we sit down, which we all did with our arms in the air, the middle and index fingers of each hand outstretched as a sign of peace. Suddenly, too, we were all intoning something that sounded part chant, part invocation and part prayer. It was soothing and melodic, and it kept repeating itself with rich, swelling rhythms. We were meeting a show of force with a show of peace.

I was sitting on the curb and there were perhaps five rows of people in front of me, separated from the police by a gap of four or five metres. The daisies I'd bought on a whim that morning were still in my hand, along with the roses the florist had insisted on giving me as a gesture of solidarity when I explained that I was part of the protest (*en solidarité*, he'd said). One by one, I passed the flowers forward, and one by one they ended up in the hands of the people closest to the police. The chanting continued, interspersed with Starhawk saying things like "We have come in peace."

A young man from our group finally stood up from where he had been sitting in the middle of the road, in front of the line of black-clad riot police. He pulled his protective bandana away from his face, then held out his hand with a rose in it, offering to give it away. Step by cautious step, he closed the distance between us and the police until finally he was almost in front of one who held a Plexiglas shield in one hand, riot club in the other. Everything was quiet except for our chanting. He held out his flower to the policeman. Nothing, though someone heard the officer mutter: *"On ne peut pas l'accepter."* He'd spoken, as a human being, telling us that he could not accept the flower. He'd replied to the gesture, in kind, and a moment of dialogue had happened in the midst of this police-state drama. The young man laid the rose on the ground at the toe of the riot cop's boot and, one by one, all the other protestors with daisies and roses went forward and did the same.

Years later, I still think about that street protest and realize that while the reports will have been long forgotten, those who experienced it will remember it always. Meaningful change will happen when it occurs inside the social and political institutions of our society, when social activists working outside government and other formal structures connect and work with those inside the centres of power, and when more and more people change the patterns of how they relate to each other and how they engage in the realities at work, at home and in society. As these people speak their minds and take the initiative, like the protestors in Port Alegre, they will gain the confidence that they can make a difference.

My good friend Ursula Franklin, now in her eighties, speaks with a wisdom that renders things simply. When we talked about how to turn things around, she said: "It's mostly a sense of priority. If you don't take your time, and really take it and say something rude to the people who try to monopolize it, it will not come."

AS I PUT my favourite bowl back on the shelf, I hope that more and more people will realize this. It's not just ourselves, our health and our relationships but our world that needs our time and attention.

Eleven

TIME FOR DIALOGUE
AND DEMOCRACY

"Caribou has run through the blood of my people for thousands of years. And I want it to endure through future generations."
NORMA KASSI, *Gwitchin Nation*, Globe and Mail

"The earth is the very quintessence of the human condition."
HANNAH ARENDT, The Human Condition

*A*S A SOCIETY, we are in the throes of a massive adaptation to the global scale, accelerated pace and multi-layered, multi-tasking complexity that restructuring, deregulation and globalization have introduced to every institution and practically every aspect of our existence in recent decades. The stress so many people feel and the symptoms of dysfunction evident in some of our social institutions, where tragic mistakes are being made because attention is shifting too much from people to data, suggest that the adaptation is not going well, at least not according to the priorities we've traditionally cherished as a society.

Yet with the acceleration of everyday life, with the instant, bit-fragmented and symbol-based medium through which so much of it is conducted, and the anaesthetizing effects of both, compliance is almost irresistible. Efficiency and its consumer corollary, instant convenience, become where it's at; the full experience of what it all means is lost in the blur.

Like all deficits, the deficit of attention and of experience adds up, with compounding interest as real bodies and relationships, real families and social institutions suffer the effects of being neglected, abandoned and deserted. It is evident among children who cannot engage effectively at school and young adults not able to think and speak for themselves. It also looms large, I have argued, in the quickening accumulation of avoidable tragedies, like baby Jordan and Walkerton, because people couldn't grasp what was really happening and act to prevent them. Our ability to function effectively as a society—to unite around collective goals and values, consenting to be governed and to govern ourselves by them because we identify ourselves as implicated participants and act accordingly at work, at home and at large in the community and the living world—hangs in the balance.

The protests at the People's Summit in Quebec City and elsewhere suggest that people sense the cracks in, if not the bankruptcy of, running society on the bottom-line rules of speed and business efficiency. There's a yearning to stop and take a critical look at the larger picture, to compare stories and to identify where and how we can intervene to negotiate the design and governance of this new space-time environment in a way that makes room for human experience and, in the process, asserts the priority of human values.

As individuals, we know that decompressing time and space wherever we can is essential, to regain a feeling for ourselves and others through fellowship and conversation, and to regain our bearings as ethical citizens. But that's only a first and personal step, and not even an easy option for those squeezed in a scramble just to survive. There remains the broader issue of the social relations of institutions, the very structures of our society and the policies behind them.

239

The fundamental question is how to get time onto the policy agenda as the medium of our social existence, not just as clocked hours of work and limits to overtime. Its absence, I think, speaks to the general inadequacy of how so much of policy-making is framed. It's so static, focussing on structures like the nation-state, the public sphere or the private, the separate integrity of which is collapsing. It's silent on the spaces in between these entities, as it is about time itself and the lessons of relativity, namely that time and space are intimately linked as the experience of life itself.

I think one way to get time on the table, and thereby to develop appropriate policies today, is by focussing less on the policy outcome, the message, and more on policy as medium. This puts time into the policy-making picture as dynamic relations between people at home and at work, within and between institutions, and inside and outside nation-states. Emphasizing policy-making as medium and process also emphasizes time as participation in dialogue, time as living bodies including living institutions and the living environment of Earth. Time as us.

THE MYTH, THE MEDIUM AND THE MESSAGE

Archetype and myth are absent from most discussions of Marshall McLuhan's famous aphorism, "the medium is the message;" yet, I think they've been there implicitly all along. Myths are seminal stories that people use to help explain their view of the world, as they serve as both currents of public thought, carrying the mind over contradictions and other obstacles, and as curbs of thought, keeping the mind within certain channels and away from others. The progress myth is a perfect example. Within myths are archetypes, original models on which others are patterned and that epitomize a concept. The archetype of the steam engine, or dynamo, as progress was literally dynamic enough to foster empires.

240

In his plays, William Shakespeare often framed choices in terms of archetypes, contrasting, for example, the royal court with the green world (often the forest). Theatrically, this approach also helped to build tension between the two competing claims to be resolved, drawing audiences into the frame of drama and dialogue. Restoring balance and equilibrium to our society in the twenty-first century requires a similar dialogue, and I think that having some mythic archetypes to think about may help us achieve that goal—but only as the beginning of broader, societal conversations, not as a stand-in featured in opinion polls and digital plebiscites.

I'd like to propose the matrix as an archetype for our times. Already, the computer-network matrix featured in the Hollywood movie by that name has become a popular symbol of cybernetic totalitarianism (genes and molecules can be rearranged in an instant because everything is pure sign, all a simulation created by computers, with nature either destroyed or enslaved as a source of fuel) and has achieved something close to the status of myth. However, as we have seen, matrix is also the cervix, which opens naturally when it's time for the baby to be born, on a cue communicated by the baby itself.

The advantage of this archetype is that it offers two world views in the one word: the space-time web of the body and, by extension, the living environment (in other words, the green world as us) plus the infinite and infinitely recombining realm of the symbol and its supporting grid of digital connections (in other words, the symbol world as us). As such, then, it also perfectly encapsulates our plight, in that we need both technology and the living environment. For both of these archetypes to co-exist we need to find balance, and the word itself reminds us that as we change one part of it, that change will automatically affect the other.

The message underlying any medium, McLuhan argued, is in how it changes the scale, speed and pattern of human affairs. As an archetype, the matrix helps to conceptualize the balance required in these three areas. The symbol matrix, rooted in the machine and whatever can be used to fuel it, is limited only by our imaginations. However, the green matrix, rooted in the world of living bodies, suggests real limits to scale-up, speed-up and complexity and to how much living things can sustain. Implicit too is the idea of environment, not as something remote from us but as something integral to us. When we consider how much stress is built into the social environment, particularly the hypermedia environment of today's commercial new world order, we can recognize the need for experience-based policy to address the problem here. Simultaneously thinking about the green world ensures that any policy-making around this new world order is hospitable to our personal health and to the health of our planet, that they are not contributing to, for example, global warming.

The next challenge is to move this dialogue forward, and to do that we need to negotiate and engage in dialogue around technology. It is strange, when you think of it, that the enormous geopolitical project of adopting standard clock-time zones as the official time around which the whole world's affairs should turn was done with no public debate or vote. This momentous decision, with all its implications for integrating transport and communication on a global scale, was treated entirely as a regulatory issue, best left to the technical experts. Still, this has been typical of discussions about technology in history. The ancient Greeks banned debates about technology from the agora because they pertained to the material world, which was the province of mere slaves and their managers, the head slave in each household. (In Plato's view, worldly mutable material things weren't worth discussing; only unchanging ideas.)

Technology jumped in status in the modern era. With people like Renaissance philosopher and statesman Francis Bacon championing a very practical approach to science and technology, it began to be understood as synonymous with progress and, once again, exempt from public debate by virtue of being above it.

To this day, technology remains largely a closed black box. It is seen as a tool for "being digital," free to use for good or bad, depending on the user or the context, which is seen as independent. There's little discussion of technology as social practices that can become rigid systems, imposing compliance. There's also little sense of technology as environment, a second nature with the power to condition adaptation, at least to some degree.

Another challenge facing this public debate will be to recognize that environments—green or technological—are not static. Dialogue will need to be continuous, as the balance is negotiated and renegotiated over time. Remember the words of Harvard biology professor Richard Lewontin: "Organisms do not experience environments. They create them... by their own activities." Urban-environmentalist Stewart Brand makes the point in *How Buildings Learn* that while buildings are the pristine products of planners and architects' designs when they're built, as they're paid off and start to run down they almost turn into putty in users' hands as an eclectic assortment of people, generally living on the margins, take over these old structures and adapt them to surprising new uses.

Former New Yorker Jane Jacobs' writings about cities offer a useful model for thinking about policy in terms of environments that are a mix of green and built worlds, where people play an active co-constituting part, formally through official jobs and informally simply as residents. Her work is helpful because she writes about cities as integrated matrices of space and time: as both a gridwork of roads and utility lines originating as symbols on a drawing board, and as the "play of public chat." To Jacobs, cities are both infrastructures and ecological habitat.

Many planners and policymakers talk about "space," but Jacobs always speaks emphatically about *place*, because time is what turns space into place, and because a city is not an abstract thing, not the way she writes about it. It's a set of intersecting relationships in a particular geographic place that exist in time, and change over time as well. These include formal relationships of business and commerce, supplemented and enriched by informal relationships in local shops and parks, on sidewalks and in the foyers of apartment buildings. If these relationships are diverse

243

enough, overlap enough in time and space (the latter through, for example, lots of short, intersecting blocks) and are able to grow over time, they produce the intangible social riches (the spontaneous conversations with shopkeepers, the petitions on store counters) that, to Jacobs, are the wealth and lifeblood of cities.

Jacobs writes with the fresh and passionate voice of an insider, what Barbara Adam, a professor of sociology at Cardiff University, calls an "implicated participant," and that's part of the message here, I think, embedded in the medium of her engaged voice as author. Her experience and the details of her daily life as a city dweller are in the picture. So she pays attention to how citizens live within the structures provided, how they bring life to them and particularly to the space-time in between the formally planned events and structures. This is the all-important space-time of unscripted dialogue, the social bridging of spontaneous civic engagement. And so when Jacobs asks what cities do, she goes beyond the usual list of services they provide such as housing and schools, beyond lights, power and other elements of the technological infrastructure. The "meat of the question," she says, is what city streets and neighbourhoods "do." They do "self-government," she says, meaning the everyday stuff by which problems are identified, priorities established and initiatives taken. At the street level, she adds, they "weave webs of public surveillance... to protect strangers as well as themselves... grow networks of small-scale, everyday public life and thus of trust and social control... and help assimilate children into reasonably responsible and tolerant city life."

From my reading of Jane Jacobs, she seems to have the same faith in people, in democracy and even in a little chaos that I do. I get a sense that she champions what is needed so that these three elements can flourish and be healthy: Above all else, discretionary time and space, decentralized decision-making and the kind of communication people need to make sense of what's happening in the constantly evolving micro-environment of the here and now, and to respond innovatively. In a word, it's dialogue.

244

DIALOGUE AS POLICY

You may have noticed that dialogue has enjoyed pride of place throughout this book. You may have thought, how romantic, how idealistic. I think it is pragmatic. Not only is conversation useful as a medium for actively slowing people down as they listen to one another, it also allows them to relate to one another on a human scale. If they are attuned to each other, dialogue models the balance of give and take that is essential to work out whatever larger issues their conversation is addressing. As an environment for communication as well as a medium, dialogue brings together the world of the body with the world of the symbol, and in a way that allows the interconnection of the two to be mediated as the discussion and the agenda evolve. When people get together to talk, in a meeting, in a classroom or hospital ward or policy committee, they can mediate the pace of communication to ensure understanding. They can interpret complexity before it becomes overwhelming. They can also say "enough" in the face of information being made accessible to them at a scale that, while nothing to a megabyte supercomputer, can bury people in minutiae, threatening their inner equilibrium and confidence in their ability to stay on top of it all.

Dialogue can help address the attention and ethical deficit in our culture, and the crisis of accountability and meaning in our social institutions in part because it's central to communicating "in good faith." Scientists talk about the "bullshit factor" as being one of evolution's more taken-for-granted gifts. It seems to arise from people's ability to read body language, to detect the tiniest flicker of an eye or eyebrow and to decipher its implications for whether someone is to be trusted or not. One enterprising researcher has identified more than eighty non-verbal elements in the face alone, and another fifty-five produced by the body. Remove the speaker from the listener, through an e-mail or a voice mail, a pulpit or a loudspeaker, and it's harder to trust. Substitute standardized data sets for conversation and it's that much harder. And if you don't know what to trust, it's harder to know what to do. Moreover, it's that much easier to be less than accountable, less than responsible and ethical. It's that much easier to just go through the motions, even doctoring the data to match "expected outcomes." After the collapse of Enron and WorldCom/MCI, Donald Johnston, secretary general of the Organization for Economic Co-operation and Development, called for a return to a culture of values and selecting independent directors based on character and "reputation." He didn't mention dialogue and a work environment conducive to implicated participation; however, it is implied that the medium that is required to express these qualities is face-to-face conversations.

Dialogue can easily become simply the flavour of the week, the buzzword of the moment on the consulting circuit, if it isn't grounded in real contexts of working things out and making decisions. To anchor it firmly at the centre of a new social contract, it must also be central to defining and negotiating accountability. This cannot be left to the technocrats; it must be negotiated among participants in the institutions subject to these accountability measures so that they feel responsible and implicated. If the message that emerges is true to the medium, it should reflect balance. It should include ways to honour the qualitative stuff of implication and the ongoing dialogue of responsible participation, not just statistical measures of production performance and service delivery.

One example of a project that involves both quantitative accountability measures and qualitative ones negotiated by both experts and participants was developed in the 1990s. An international development project based in Bangladesh required an evaluation technique that would indicate the "success" of the group's actions despite outcomes that sometimes defied quantification and prediction. How to measure this? The stakeholders in this project—and there were many—pioneered "most significant change" (MSC). It's an evaluation technique based on dialogue and story that involves multiple levels of scrutiny yet hints at fruitful change and promising new directions for assistance.

This "dynamic values inquiry" begins with storytelling. Once a funded program is complete, participants are each asked to say what they think has been the most significant change it has caused in their lives and to explain why. Next, the stories are interpreted by the group and its leaders in light of the expected outcomes and goals of the program.

From this dialogue, "winning" stories (what the local group considers the most meaningful change) are sent, as part of the reports, to a higher level of authority, often the managers of the non-governmental organization (NGO) that provided the original funding. The managers select a smaller number of winning stories, again from a dialogue addressing the all-important question of why the changes were significant, that best represent in their language of systems and administration the goals they were hoping to achieve.

While the winning stories from this round move up to, for instance, an outside funding agency, such as the Canadian International Development Agency or the World Bank, for further selection according to that body's language and perspectives, the results of the NGO managerial selection are relayed back to the organization and its program participants.

Although MSC is a time-consuming exercise, when combined with more traditional evaluations using predetermined quantitative indicators it helps an organization "adjust the direction of its attention." As such, it's been called an "evolutionary epistemology"—which is fitting, I think, considering new insights into how even genetic mutations emerge as organisms respond to changing conditions in their environment and, through their adaptations, fine-tune their fit as contributing parts of it.

MODELS OF CHANGE

The example of MSC demonstrates that dialogue is useful not just as a tool for talking about things but also as a principle for negotiating policy, and this is its virtue as a linchpin of a new social contract, balancing the competing claims of the symbol world and the green world in our lives and by implication our social institutions. Though I can't predict a whole policy agenda arising from this concept, some promising hints of this change are beginning to show. An initiative by a group of *Karoshi* widows, wives of employees killed by overwork, is perhaps a case in point here. They are lobbying the Japanese government to enshrine livable working hours as central to people's basic human rights. For them, the right to work in an environment that is hospitable to people's health and longevity is an extension of people's basic right to exist. By implication, this right to exist must include the right to a living wage, that is, the right to a rate of compensation for one's time that will allow one to make ends meet and have a life, too.

In Britain, the same argument has been made through the courts. Lawyers for an overworked social worker who suffered a second nervous breakdown after his employers failed to reduce the stress and workload that had contributed to the first one successfully argued that a combination of rising work demands and lack of personal control and working guidelines did "psychiatric damage to a normally robust personality." The claimant won a large compensation package.

I've noticed some changes in policy language that give me hope. For example, in a legal position paper prepared by the Department of Foreign Affairs and International Trade accompanying an amendment to the International Boundary Waters Treaty Act to prohibit bulk exports of water, the authors championed the need to preserve "the integrity of the ecosystem within water basins" and with this, the difference between water as a commodity and a living body.

The Kyoto Protocol on the environment is also an encouraging sign, but it doesn't seem to have any teeth or much buy-in, even from the signatories. So how do we ensure that dialogue takes place effectively and that one entity, like the transnational business community, doesn't just impose its will on people? The key, I

think, is expanding the scope and the time for democratic participation, both as dialogue and action. We will make the priorities of
our green world real as we talk about the stress we as individuals,
as families and as social institutions are under, and sense the parallels to a living earth that's overextended and under stress. We will
gain the will to speak and act for policies such as the Kyoto Protocol as we feel the legitimacy and urgency of speaking and acting on
the more personal level, and become aware of the interconnections between the two.

We are used to governments and the courts making decisions
for us; however, the realities of a more fluid, complex and interde-
pendent world demand the flexibility of a more participatory
model. Jane Jacobs has recently suggested that cities adopt "performance codes" on issues such as noise and density, yet leave it to
people and businesses that locate there to figure out how they will
achieve these agreed-upon public-interest goals. This idea could
be usefully generalized in a new approach to governance locally
and globally, inside institutions and between them. I envisage
some formal policy-making and negotiation to establish common
standards and goals, like the global emission-reduction standards
associated with the Kyoto Protocol and national standards in
health care. The actual strategies for achieving the goals account-
ably could be worked out locally, within institutions, within and
between working teams, given enough discretionary time and
space to achieve this. As we implicate ourselves in negotiating responsible strategies, our cities, public buildings and institutions in
turn might better reflect the goals we as a society have set for
them, in ways that make sense to the people involved.

There are some encouraging initiatives in institutional self-
governance, both in the private sector (in information sharing and
"mutual gains" negotiating between workers and management, in
some pulp-and-paper companies, for example) and in the public
sector. A study of 220 school boards across North America ranked
Edmonton, Alberta, public schools among the top six not just

because their students perform well on tests but because the schools have created a community environment that invites students to buy into schooling. One key here, the study found, was local school autonomy through school-based budgeting. Principals are free to interpret broad curriculum goals and rules in ways that best reflect the circumstances of the local school population and respond to changing trends. Two schools in East Harlem, New York, have copied Edmonton's strategy, with startling success. Deborah Meier, the principal at one of those schools, is credited with creating a genuine community in her inner-city school by budgeting based on priorities identified by the students themselves. She has emphasized spending money on small class sizes and on retaining teachers through multiple-year assignments. With more stability and continuity for students as they move through the grades, the dropout rate has plummeted, and 90 per cent of students at these East Harlem schools go on to college.

The dialogue model as policy and policy-making medium embedded in the living context could bring the personal politics of social movements together with the more formal geopolitics of the globalized world. In doing so, it can bridge the gap between what makes sense to us as technocrats (operating on symbol-sphere logics of efficiency and speed) and what makes sense to us as parents of a child with attention deficit disorder, or as the offspring of a parent with dementia treatable only with care or simply as people who want a life and don't want to crash with a heart attack or chronic fatigue. The policy agenda would need to address what is necessary for effective dialogue, including the discretionary authority to take one's time to nurture healing and preventive-health-care relationships, learning and mentoring in the educational sphere, agency for change in social services, and a healthy work-and-life balance everywhere. It might also highlight the time and pace needed for reflection, interpretation and consensus-building in running social institutions through semi-autonomous,

self-governing teams that, while accountable to larger performance codes, standards and objectives, are left to work out how best
to achieve these under changing circumstances. Institution by institution, the dialogic model could restore the time needed for
effective deliberative and participatory democracy. It might also
restore the dream of a leisure society, by cultivating a taste for a
more leisurely pace in everything.

We are the experts we've been waiting for.

EPILOGUE

"Peace is not just the absence of war. It is the presence of justice."
URSULA FRANKLIN, CBC Ideas

*"The ideal subject of totalitarian rule is not the convinced Nazi
or the convinced Communist, but people for whom the distinction
between fact and fiction (i.e., the reality of experience) and the
distinction between true and false... no longer exist."*
HANNAH ARENDT, The Origins of Totalitarianism

SEPTEMBER 11, 2001, came and went while I was working on this book, and though I sensed that it was significant to what I was writing I only now understand why. For me, the attacks on the World Trade Center and their aftermath drive home the need for dialogue in an interconnected world even as they threaten to silence it. They also crystallize the choices we face, particularly the two fundamental ones I've discussed: how we relate to and engage with one another in this globalized world, and how the realities we act on will be defined and articulated.

Both the attacks and the U.S.-led "war on terrorism" that followed signal a dangerous turn. They mark not only the near supremacy of the symbol sphere (which is dominated by sound bites, data and iconic images) over nature and society, but its potential to enclose us in an undemocratic new world order that has us perpetually in a state of war or war preparedness, in the face of which face-to-face diplomacy and reconciliation can have at best a merely mitigating effect. I feel that this could happen not just because of how information is controlled or because updates and new developments unfold at a pace that outstrip sober second thought; it could also occur because too many of us, individually and collectively through our social institutions, have lost the ability to think and speak for ourselves, and have given up the time and the opportunities for dialogue in shared time and space that this requires.

Those who depict the "war on terrorism" as a choice between "McWorld" and "Jihad" aptly name the clash of values between commercialism and spiritualism, but they miss a larger message. Not only do both choices compress a complex interplay of economic, political and spiritual factors into a dichotomy, both are cast in the medium of symbols, or at least they were on September 11. Saudi-born dissident Osama bin Laden is almost a video-game avatar of Jihad, while U.S. president George W. Bush, with his friends in big oil and big business, is the face of McWorld.

There was no more obvious symbol of McWorld for bin Laden and his al-Qaeda faction to target than the larger-than-life twin towers in New York City's financial district, where trillions of dollars in global investments turn over every day. It isn't known whether the two-step timing of the attacks was deliberate, but it couldn't have been better scripted for symbolic effect. The first jet that struck the towers cued the cameras for the second and the assault unfolded as a televised spectacle: the epitome of corporate global power first being summarily breached and then crashing to the ground.

I can still remember the jolt, jolt, jolt I felt as I tuned into the television reports that morning and watched the repeated image of the second passenger plane hitting the second tower and, later, the hollow, sick feeling I had again and again as the TV replayed the sight of the North Tower suddenly disintegrating into a maelstrom of dust and debris. Equally, I remember how quickly interpretation of the attacks was abbreviated into "evil," "terrorism" and "war." I sat in my chair in front of the TV, too numb to even wonder at the inappropriateness of the terms. Within twenty-four hours of the attacks, the images on the screen and the powerfully evocative yet largely rootless words accompanying them had turned the most powerful country in the world, the United States of America, into a fortress, with personal toiletry items reclassified as weapons. The PATRIOT (Providing Appropriate Tools Required to Intercept and Obstruct Terrorism) Act passed Congress in late October, granting sweeping powers of arrest, detention, deportation and extradition without judicial oversight. The Homeland Security Act and Information Awareness Office extended these powers. Privacy and civil liberties were peremptorily sacrificed to the goal of identifying any potential threat, using the latest surveillance techniques to scan Internet and cellphone traffic, plus video images of pedestrian traffic at airports, malls, hotels and hospitals to detect any suspicious patterns.

Data are now being collected on our every movement, and these are being interpreted through a filter of suspicion, with the result that data profiles can eclipse, and compromise, the full reality of who people actually are.

And so, what should have been a healthy, democratic discussion about the meaning of the World Trade Center attacks in a world of increasing inequalities and a worrisome concentration of power into the hands of a few mainly U.S.-based transnational corporations, and what could perhaps have brought about some peace and reconciliation initiatives to minimize public sympathy for terrorism by demonstrating that more reasoned, deliberative alternatives do work, has instead become a rigid, crippled silence. People are afraid to speak out lest they be identified as a "terrorist" or a "terrorist supporter." Political leaders are afraid to stray from safe, well-tested positions lest, in daring to engage in a nuanced public debate, they prompt talk-show hosts to turn to a guest more reassuringly associated with sponsors and the status quo, or viewers to simply click to a faster-paced channel.

I became more skeptical about media sound and sight bites when the U.S.-led "coalition of the willing" invaded Iraq in March 2003 (ostensibly because its regime under the "brutal" Saddam Hussein was linked to terrorism, with alleged weapons of mass destruction aggravating the threat). I became even more so when American reporters were "embedded" into armoured vehicles. From that closed-in perspective they had no choice but to project the U.S. war machine advancing triumphantly toward Baghdad and fulfilling its goal, virtually and symbolically, by toppling the colossal statue of Hussein, the country's military leader, in the centre of the city. According to reports that circulated, unverified and unverifiable, on the Net, the real Hussein was first found by Kurdish troops before being officially captured in a code-named American operation broadcast on American television. When the Arab television station Al-Jazeera broadcast the decapitation of an American contract worker several months after the initial invasion, some viewers questioned the uneven time codes on the video, suggesting that media manipulation is not the domain of the Americans alone.

"I don't know what to believe any more," a friend lamented over dinner. Here I sense the danger of paralyzed passivity in the face of fast and facile symbols and data that can readily be doctored because they are devoid of context and far removed in both space and time from realities we can verify or relate to. I worry that when a society experiences mass helplessness and uncertainty, the certainties of fascist-like control can become irresistibly attractive. Political theorist Hannah Arendt, writing about the rise of Nazism in Germany and Communism in Russia, found that these totalitarian states succeeded because traditional forms of public consensus-building and civic dialogue had broken down. It's when "common sense has lost its validity," she argued, that totalitarianism can succeed with its outrageous logic. Abstractions, stock phrases, slogans and other elements of "officialese" fill the gap with their own entirely symbol-based reality. The regime (and its symbols) becomes a law unto itself, answerable only to the logic of its will and its pace of constantly changing, constantly updated developments.

We can sleepwalk into this state as both a daily regime and a geopolitical one if we let ourselves become too stressed and anaesthetized by our busy, busy lives to focus on anything but what's on the screen or our daily to-do list, or if we become too distracted by celebrity TV, the numbers on the stock market, the bargains on eBay or at the local mall. We can drift into totalitarianism almost ineluctably because the reality brought to us by statistical indicators or larger-than-life icons is what we're used to, or because we've forgotten how to pay attention, to slow down and focus on our own thoughts, values and experience instead of just going with the flow.

I would argue that many people have already grown comfortable with, even dependent on, official statements of reality, for example that "we" are winning the "war on terrorism," even when the more qualitative, experience-based evidence suggests that the world is becoming more dangerous and retreating from the values that the war is purportedly defending. The habits of dialogue and self-governance, which can determine the priorities that are important to us as individuals and as a society (and which we can make real by acting on them) are withering, and the necessary social skills plus the feeling for its slowed, more engaged pace, are atrophying. But it is not too late. We can still choose to be in touch with ourselves and to engage directly in the realities of our world and with each other.

I am struck by the fact that while the symbols were unfolding on the screen that terrible day in September 2001, another reality was unfolding within the twin towers themselves, an incredible dialogue of survival. Although some people followed the directive issued over the public address system that everything was okay and they could go back upstairs to work, most paid attention to the realities around them. They helped each other down the stairs and were led to safety by the selfless efforts of the New York firefighters and police officers. Ninety-nine per cent of the people on the floors beneath where the planes hit made it out alive that day. According to one survivor's report, "People were fainting, collapsing, being passed along overhead so they wouldn't slow down the escape too much." Cracks appeared in the walls, steam pipes burst, emergency phones didn't work, though the sprinklers did, sending water cascading down the stairs amid the dust and the smoke.

Another survivor wrote: "When fears bubbled up, there were always reassuring words from the right person that calmed even the most frightened of us.... Since September 11 I've talked to many people who were inside that day, and many more who watched and waited. Each person I meet has a story to tell, and an urgency to press the flesh. It's important to reassure oneself that the person in front of them is real. It's also important for me personally to get the right message out. In the stairwell, people treated one another so beautifully. No one stopped to think what colour or religion the other human being was... "

I was in Ottawa that day. All over the city, people poured out onto the streets, hugging each other, trading news, thoughts, feelings. Hundreds, possibly thousands, went to the fortress-like U.S. embassy and completely covered the row of thick iron bars in front of it with flowers, stuffed animals and cards, in French and in English, expressing sympathy, condolences, only rarely anger or vengeance. Thousands lined up at emergency blood-donor clinics set up that evening. One donor, who signed his name only as "Walter," left on the counter a note he'd written while waiting in line: "Today I saw humanity in the face of inhumanity. In the midst of carnage, I witnessed the most powerful force on earth. Not the bomb or the military or an unspeakable act of terror. Today I was surrounded by mass compassion...the givers of life, blood donors." For me, those actions were proof that as citizens we are still able to engage with the people and the realities around us.

257

As the immediate crisis died down, North American sales of books about Islam and the Middle East increased, which suggested that citizens wanted to learn more about the history, politics and culture of the people being villified as "terrorists." As well, a host of spontaneous initiatives in interfaith dialogue and cross-cultural community-building have occurred, including a Kids4Peace project that brought together Muslim, Jewish and Christian children from the Middle East and Canada first in correspondence and then in a summer camp so they could learn about each other.

I want to see in these efforts a hopeful awakening, a ground-swell of recommitment to the values and, more to the point, the time and social habits on which peace in our lives and peace in our world depend. But to democratize the new world order and direct it to serve peace and equilibrium requires more than a few well-intentioned gestures. More and more of us must break out of our isolated individualism and distracted busyness and join the dialogue that is society. September 11 and what followed have raised the stakes. Being elsewhere, being otherwise engaged, is no longer an option.

NOTES

INTRODUCTION

p2 *space-time compression* David Harvey, *The Condition of Postmodernity* (Cambridge, MA: Blackwell, 1989), especially Parts II and III.

p3 *"battle stress"* Terry Copp and Bill McAndrew, *Battle Exhaustion* (Montreal: McGill–Queen's University Press, 1990), 22.

p3 *a kind of anaesthetization* Hans Selye, *The Stress of Life* (Toronto: McGraw Hill, 1956), 264. Although Selye uses the word "anesthesia," the emphasis I give the term is largely mine as I draw together what scientists have learned and interpret it broadly as human experience, much as Marshall McLuhan emphasized the numbing effects of stress in drawing on Selye's work.

p3 *coherence and equilibrium* Hans Selye, *Stress without Distress* (Toronto: McClelland & Stewart, 1974), 35.

p3 *"sleeping-brain dialogue"* Jeremy Campbell, *Winston Churchill's Afternoon Nap* (New York: Simon & Schuster, 1986), 182.

p8 *A U.S. presidential administration feeds false* John MacArthur, "All the News that's Fudged to Print," *Globe and Mail,* 6 June 2003, A11.

p9 *almost routinely massaged* Eric Reguly, "Avert Future Enrons with More Audit Independence," *Globe and Mail,* 5 February 2002, B16. For a more comprehensive review see David Henry, "The Numbers Game," *Business-Week,* 14 May 2001, 100–04.

p9 *EBS* Henry, 103.

p9 *"fast cycle times"* James Gleick, *Faster: The Acceleration of Just About Everything* (New York: Pantheon Books, 1999), 75.

p9 *"The Cult of Efficiency"* Janice Gross Stein, *The Cult of Efficiency* (Toronto: Anansi Press, 2001).

p13 *memories of having done this* George Painter, *Marcel Proust: A Biography* (London: Penguin Books, 1983), 448–49.

CHAPTER 1: BUILDING AN ENVIRONMENT IN MOTION

I read some twenty books on the general subject of time. My favourites, not cited elsewhere here, include: Johannes Fabian, *Time and the Other: How Anthropology Makes its Object* (New York: Columbia University Press, 1983); Francis Fukuyama, *The Great Disruption: Human Nature and the Reconstitution of Social Order* (New York: Free Press, 1999); Jacob Needeman, *Time and the Soul* (New York: Currency/Doubleday, 1998); Helga Notwotny, *Time: The Modern and Postmodern Experience* (Cambridge: Polity Press, 1994). I also trust almost anything by Barbara Adam and Nigel Thrift, and the journal *Time & Society*.

p18 *"highways of seawater"* Homer, *The Odyssey,* trans. Robert Fitzgerald (Garden City, NY: Doubleday, 1961), 49.

p18 *fifteen busiest ports* Jeremy Relph, *A Quick Guide to Global Container Hubs* (Toronto: Canadian Transportation and Logistics, June 2000). (Also available on-line: www.ctl.ca/research/marine_carriers/features/global_container_ hubs.asp)

p19 *lions were transported* Eric Rath, *Container Systems* (New York: John Wiley & Sons, 1973), 6.

p20 *"signified or expressed man"* Armand Mattelart, *The Invention of Communication* (Minneapolis: University of Minnesota Press, 1996), 40.

p20 *kilogram standard* Mattelart, 42.

p21 *"We are time"* Barbara Adam, *Timewatch: The Social Analysis of Time* (London: Polity Press, 1995), 145.

p21 *rhythm in relationships* Edward Hall, *The Dance of Life: The Other Dimension of Time* (Garden City, NY: Doubleday, 1984), 153. Hall writes: "Rhythm is the essence of time."

p22 *These annual cycles* Nigel Thrift, "Owners' Time and Own Time: The Making of a Capitalist Time Consciousness, 1300–1880," in *Space and Time in Geography* (Lund, Sweden: CWK Gleerup, 1981), 58.

p22 *"disassociated time"* Lewis Mumford, *Technics and Civilization* (New York: Harcourt, Brace & World, 1963), 15.

p23 *"medium of existence"* Mumford, 17.

p23 *Newton's notions* Barbara Adam, *Timescapes of Modernity: The Environment and Invisible Hazards* (London: Routledge, 1998), 41.

p23 *"industry of every kind"* Adam Smith, *An Inquiry into the Nature and Causes of the Wealth of Nations* (New York: Random House, 1937), 18.

p24 *time-motion studies* Daniel Nelson, *Frederick W. Taylor and the Rise of Scientific Management* (Madison: University of Wisconsin Press, 1980), 41.

p24 *"signatures"* Art Shaw, historian, personal communication.

p25 *Speedee Service System* Eric Schlosser, *Fast Food Nation: The Dark Side of the All-American Meal* (New York: Houghton Mifflin, 2001), 20.

p26 *"aura"* Walter Benjamin, *Illuminations* (London: Fontana/Collins, 1979), 222–23.

p27 *"space of flows"* Manuel Castell, *The Informational City: A New Framework for Social Change* (Toronto: Centre for Urban and Community Studies, University of Toronto, 1991), 14.

p28 *Sir Sandford Fleming* Clark Blaise, *Time Lord* (Toronto: Vintage Canada, 2001).

p28 *"kind of hyper time"* Blaise, 129.

p28 *"terrestrial time"* Blaise, 80.

p28 *"cosmic time"* Blaise, 128.

p28 *"We are now obliged"* Blaise, 35.

p29 *International Organization for Standardization* C.Douglas Woodward, *BSI: The Story of Standards* (London: British Standards Institution, 1972), 62.

p30 *$2 trillion changes hands* Ronald Deibert, *Parchment, Printing, and Hypermedia: Communication in World Order Transformation* (New York: Columbia University Press, 1997), 151.

p30 *connecting grammar* Jim Kerstetter, "When Machines Chat," *BusinessWeek,* 23 July 2001, 76.

p32 *auditing is used* Judy Hunter, "Working Life & Literacies at the Urban Hotel," in *Reading Work: Literacies in the New Workplace,* Mary Ellen Belfiore et al. (Mahwah, NJ: Lawrence Erlbaum Associates, Publishers, 2004), 101–48.

p33 *education as standardized deliverables* Marjorie Griffin Cohen, "Trading Away the Public System: The WTO and Post-secondary Education," in *The Corporate Campus: Commercialization and the Dangers to Canada's Colleges and Universities,* James Turk, ed. (Toronto: James Lorimer, 2000), 132.

p35 *Wayne's trucker buddies* Most of the comments included here come from a separate interview I did with a group of truckers in February 2001, which was arranged by Dave Tilley of the Canadian Auto Workers.

p36 *global-warming carbon dioxide* ICF Consulting, *North American Trade and Transportation Corridors: Environmental Impacts and Mitigation Strategies,* 21 February 2001, 11.

p38 *something intimate, fiduciary* James Carey, *Communication as Culture* (Boston: Unwin Hyman, 1989), 18.

p39 *collapsing the difference* Deibert, 210.

p39 *"segments defined in terms of"* Deibert, 184.

p40 *"'message' of any medium"* Marshall McLuhan, *Understanding Media: Extensions of Man* (Cambridge, MA: MIT Press, 1999), 8.

p40 *"psychic and social consequences"* McLuhan, 11.

p41 *fast fashions* Karen von Hahn, "Noticed: Fast Fashion," *Globe and Mail,*
    6 March 2004, L3.

p43 *fast drives out the slow* Thomas H. Eriksen, *Tyranny of the Moment: Fast and
    Slow Time in the Information Age* (Sterling, VA: Pluto Press, 2001), 127.

p44 *"globalism"* John Ralston Saul, "The Collapse of Globalism" *Harper's,*
    March 2004, 33.

p45 *trucks hauled more than 61 million* ICF Consulting, 1.

p47 *falsified logbooks* Gary Dimmock and Zev Singer, "McCain: 1 Million
    Pounds of Fries an Hour," *Ottawa Citizen,* 23 June 2001, A1. In another arti-
    cle (6 September 2001, A1), the same writers reported that the Canadian
    Automobile Association estimated that tractor-trailers and other heavy
    trucks were involved in 12.5 per cent of fatal crashes in Ontario, although
    they represented less than 3 per cent of total vehicles on the road.

CHAPTER 2: STRESSED OUT AND DREAMLESS

p51 *dispersed subject* Mark Poster, *The Mode of Information: Poststructuralism
    and Social Context* (Cambridge: Polity Press, 1990), 123.

p52 *working more than forty-five hours* Linda Duxbury and Chris Higgins, *Work-
    Life Balance in the New Millennium: Where are We? Where Do We Need to Go?*
    (Ottawa: Canadian Policy Research Network, 2001), 19.

p52 *"highly stressed"* Duxbury and Higgins, 16.

p53 *"lifeline"* David Rocks, "The Net as a Lifeline," *BusinessWeek e.biz,*
    29 October 2001, 16.

p54 *"iron law of wages"* William Pfaff, "The Price of Globalization," *International
    Herald Tribune,* 10 January 2004, available on-line:
    www.iht.com/articles/124402.html.

p54 *"factories for hire"* Janet McFarland, "Factories-for-hire Feed High-tech
    Assembly Boom," *Globe and Mail,* 8 February 1996, B12.

p55 *the real competition* Alan Freeman, "Is this a Local Call? Not Exactly," *Globe
    and Mail,* 23 March 2002, F6.

p55 *40 per cent were part-time* John Dillon, "Canada Has Not Enjoyed the
    Benefits of NAFTA Its Advocates Promised," *CCPA Monitor* (November
    2003), 47.

p55 *8.1 millon Canadians* Duncan Cameron, "Canada's Social Activists Should
    Mount a 'Living Wage' Movement," *CCPA Monitor* (December 2003/January
    2004), 16.

p56 *four hours a week* Duxbury and Higgins, 19.

p56 *they took on extra work* Rashida Dhooma, "Hey, Boss, Is It Nap Time Yet?"
    Toronto Sun, 9 April 1998, 74.

p56 *increased demands and expectations* Heather Menzies and Janice Newson, "Academic Time Survey" (unpublished).

p56 *"mobile professionals"* James Gleick, *Faster: The Acceleration of Just About Everything* (New York: Pantheon Books, 1999), 91.

p56 *majority of home computers* Heather Menzies, *Women and the Knowledge-Based Economy and Society* (Ottawa: Status of Women Canada, 1998), 11. The statscan.ca Web site also reports that in 1999 one in five households used their home Internet connection for self-employment, and one in four did so for employer-related reasons.

p57 *"one-dayers"* Michael Young, *The Metronomic Society: Natural Rhythms and Human Timetables* (Cambridge, MA: Harvard University Press, 1988), 217.

p57 *"move on"* Sarah Koch-Schulte, "Cheeky Operators: Resistance Tactics in Canada's Call Centres," in *Just Doing it: Popular Collective Action in the Americas*, G. Desfor et al., eds. (Montreal: Black Rose Books, 2002), 163.

p58 *Dark Age* Jane Jacobs, *Dark Age Ahead* (Toronto: Random House of Canada, 2004).

p58 *three times more likely* Jane Coutts, "Workplace Stress More Prevalent than Illness, Injury," *Globe and Mail*, 8 April 1998, A2.

p59 *a chronic disease* K.K. Campbell, "Electronic Snoops Are on the Job," *Toronto Star*, 19 March 1998, J1.

p59 *overwhelmed by stress* Sean Fine and Marian Stinson, "Stress Is Overwhelming People: Study," *Globe and Mail*, 3 February 2000, A1.

p59 *10,000 deaths* Tetsuro Kato, "The Political Economy of Japanese Karoshi (Death from Overwork)," *Hitotsubashi Journal of Social Studies* 26 (1994), 44.

p59 *workload and deadlines* Kensington Technology Group study, San Mateo, CA, reported in a Knight-Ridder wire story, *Ottawa Citizen*, 15 August 1998, J2.

p59 *a third of U.S. workers* Diane Stafford, "Stressed-out: Nearly One-third of U.S. Workers Say They Feel Overwhelmed on the Job," *Montreal Gazette*, 19 May 2001, C9.

p59 *"family-time"* Linda Duxbury and Chris Higgins, *Report One: The 2001 National Work-Life Conflict Study* (Ottawa: Health Canada, 2002). (Also available on-line: www.phac-aspc.gc.ca/publicat/work-travail/index.html.)

p59–60 *survey of social-services workers* CUPE Research Branch, *Overloaded and Under Fire: Municipal Social Services Workers* (Ottawa: Canadian Union of Public Employees, 2001), 1.

p60 *Work-related learning* Duxbury and Higgins, *Report One: The 2001 Work-Life Conflict Study*.

p60 *decision-making is centralized* Wayne Cornell and Louise Lemyre, *The Health Status of Executives in the Public Service of Canada, Preliminary*

Findings (Ottawa: Association of Professional Executives of the Public Service Alliance of Canada, November 2002). (Also available on-line: www.apex.gc.ca.)

p60 *lack of supportive supervisors* Michael Kesterton, "Long Hours No Problem?" *Globe and Mail*, 30 October 2003, A18.

p61 *good manners* Grant Buckler, "Companies Battle Rudeness," *Globe and Mail*, 19 November 2003, C1.

p62 *"auto-amputations"* Marshall McLuhan, *Understanding Media, Extensions of Man* (Cambridge, MA: MIT Press, 1999), 43.

p63 *define the boundaries* Rohan Samarajiva, "Privacy in Electronic Public Space: Emerging Issues," *Canadian Journal of Communication*, Vol. 19 (1994), 90.

p63 *A British study reported* Jane Wills, "Laboring for Love? A Comment on Academics and their Hours of Work?" *Antipode* 28:3 (1996), 295.

p64 *cost in health-related absenteeism* Linda Duxbury, Chris Higgins and Karen Johnson, *Report Three: Exploring the Link between Work-Life Conflict and Demands on Canada's Health-care System* (Ottawa: Health Canada, 2004).

p64 *second most-prescribed* Mark Kingwell, *Better Living: In Pursuit of Happiness from Plato to Prozac* (Toronto: Viking, 1998), 96. Statistics kept by the Health Information Institute are more cautious, though they continue to list this class of drugs among the top five, along with heart and stomach medications to relieve stress and its symptoms.

p64 *perfect weekend pick-me-up* Danylo Hawaleshka, "Viagra's New Competition," *Maclean's*, 15 December 2003, 36.

p64 *disability and death* "Overload Can Kill You, Family Institute Warns," *Montreal Gazette*, 19 March 2001, E2.

p64 *1.5 million Canadians* Duxbury and Higgins, *Report One: The 2001 Work-Life Conflict Study*, 15.

p64 *33 per cent of working Canadians* Duxbury and Higgins, *Report One: The 2001 Work-Life Conflict Study*.

p65 *caffeine jolt* André Picard, "Tests Find Java Packs Huge Jolt" *Globe and Mail*, 19 May 2001, A1.

p66 The Lancet *described* Wolfgang Schivelbusch, *The Railway Journey: The Industrialization of Time and Space in the 19th Century* (Berkeley: University of California Press, 1986), 56.

p66 *"shell shock"* Terry Copp and Bill McAndrew, *Battle Exhaustion*, (Montreal: McGill–Queen's University Press, 1990), 22. For a more extensive discussion, including conditions during the Vietnam War, see Albert Glass's "Introduction" to *The Psychology and Physiology of Stress* (New York: Academic Press, 1969), xxiii.

p67 *rotate people away* Glass, xx.

p67 *"sustaining relationships"* Glass, xix.

p67 *necessary period of calm* Archibald D. Hart, *The Hidden Link between Adrenalin & Stress: The Exciting New Breakthrough that Helps You Overcome Stress and Damage* (Waco, TX: Word Books, 1986), 51.

p67 *"adaptation energy"* Hans Selye, *Stress without Distress* (Toronto: McClelland & Stewart, 1974), 38. For a good discussion of the phases of stress and Selye's theory of general adaptation, see *The Stress of Life* (Toronto: McGraw-Hill, 1956), 31.

p68 *"essence of the stress problem"* Hart, 49.

p68 *perceived loss of control* Sonia Lupien, "Stress-function-morphology Correlations in the Brain," *McGraw-Hill Yearbook of Science and Technology* (New York: McGraw-Hill, 2003), 418.

p68 *stomach medications* Alain Vinet et al., "Piecework, Repetitive Work and Medicine Use in the Clothing Industry," *Soc. Sci. Med.*, Vol. 28, No. 12 (1989), 1287.

p68 *and tranquilizers* Herbert Northcott and Graham Lowe, "The Influence of Working Conditions on Psychological Distress in the Post Office," *Canada's Mental Health* (September 1984), 26.

p68 *"burden on the cardiovascular system"* Takeshi Hayashi et al., "Effect of Overtime Work on 24-hour Ambulatory Blood Pressure," *Journal of Occupational & Environmental Medicine* Vol. 38, No. 10 (October 1996), 1010.

p68 *serum cholesterol* Robert Karasek and Tores Theorell, *Healthy Work: Stress, Productivity and the Reconstruction of Working Life* (New York: Basic Books, 1992), 147.

p68 *"mind becomes obsessed"* Hart, 81.

p68 *salivary immunoglobulin A* Hans Zeier et al., "Effects of Work Demands on Immunoglobulin A and Cortisol in Air Traffic Controllers," *Biological Psychology* 42 (1996), 414.

p68 *"control is difficult"* G. Robert Hockey et al., "Intra-individual Patterns of Hormonal and Affective Adaptation to Work Demands: An n=2 Study of Junior Doctors" *Biological Psychology* 42 (1996), 395.

p69 *physiological rhythms* Robert Henning and Steven Sauter, "Work-physiological Synchronization as a Determinant of Performance in Repetitive Computer Work," *Biological Psychology* 42 (1996), 281.

p69 *atrophy the brain's hippocampus* Lupien, 418. For more detail, see also her article "The Neuroendocrinology of Cognitive Disorders," in *Biological Psychiatry* (2 vols.), H. D'haenen, ed., (New York: John Wiley & Sons, 2002).

p69 *Cortisol suppresses* Zeier et al., 414.

p69 *linked to stress* Gabor Maté, *When the Body Says No: The Cost of Hidden Stress* (Toronto: Alfred A. Knopf, 2003), asthma, 90; Alzheimer's, 162; rheumatoid arthritis, 177.

p69 *"don't need much sleep"* Peter Martin, "Stolen Slumber," *Toronto Star*, 7 February 1999, F1.

p69 *"sleep stupidity"* Stanley Coren in Martin, F1.

p70 *"dangerously sleep-deprived"* Stanley Coren in Jennifer Hunter, "Are You Getting Enough Sleep?" *Maclean's*, 17 April 2000, 43.

p70 *National Sleep Foundation* Nancy Ann Jeffrey, "Sleep: The New Status Symbol," *Globe and Mail*, 10 April 1999, D5.

p70 *time survey* Heather Menzies and Janice Newson, "Academic Time Survey" (unpublished).

p70 *"slow wave" deep sleep* Martin, F5.

p71 *"sleeping-brain dialogue"* Martin, F5.

p71 *our nearest primate relatives* Stanley Coren, *Sleep Thieves* (New York: Free Press, 1997), 249.

p71 *"hypnagogic half-dreaming"* Martin, F5.

p71 *low-adrenalin arousal* Hart, 196.

p71 *Dream time* Bruce Chatwin, *The Songlines* (London: Penguin Books, 1988), 2.

p71 *"trail on earth"* Robin Ridington, "From Artifice to Artifact: Stages in the Industrialization of a Northern Hunting People," *Journal of Canadian Studies* Vol. 18, No. 3 (Fall 1983), 59.

p72 *Technology as Symptom* Robert Romanyshyn, *Technology as Symptom and Dream* (New York: Routledge, 1989). See also his article "The Dream Body in Cyberspace," *Psychological Perspectives*, Issue 29 (1994).

CHAPTER 3: WORKAHOLICS AND CHRONIC FATIGUE

p79 *classic narcissists* Barbara Killinger, *Workaholics: The Respectable Addicts* (Toronto: Key Porter Books, 1991), 38.

p79 *dissociated from their feelings* Killinger, 56.

p79 *over-plan and over-control* Killinger, 65, 89.

p79 *tunnel vision* Anne Wilson Schaef, *When Society Becomes an Addict* (San Francisco: Harper & Row, 1987), 86.

p79 *cut themselves off* Killinger, 71.

p79 *out of touch* Killinger, 127.

p79 *often on-line* There's a wealth of research on this, some focussed on cybersex addiction. For more general comment, see Anne-Marie Owens, "If They Live Online, Are They Addicts?" *National Post*, 3 March 2001, A1.

p79 *compulsive need* Killinger, 135.

p80 *"We live in a workaholic society"* Personal conversation with Dr. Killinger, Toronto, June 2002.

p80 *"This extension of himself"* Marshall McLuhan, *Understanding Media, Extensions of Man* (Cambridge, MA: MIT Press, 1999), 41.

p83 *stats on attention deficit disorder* "Kiddie Coke," *Globe and Mail,* 7 April 2001.

p84 *rapid change involving "dislocation"* Bruce Alexander, *The Roots of Addiction in Free Market Society* (Ottawa: Canadian Centre for Policy Alternatives, 2001), 3.

p85 *time as continuity* Gerda Reith, "In Search of Lost Time: Recall, Projection and the Phenomenology of Addiction," *Time & Society,* Vol. 8(1) (London: Sage, 1999), 101–03.

p85 *"entrainment, or mode locking"* James Gleick, *Chaos: Making a New Science* (New York: Penguin, 1987), 292.

p86 *"temporal fingerprint"* Robert Levine, *A Geography of Time* (New York: Basic Books, 1997), xvii.

p86 *"enables us to move together"* Jeremy Campbell, *Winston Churchill's Afternoon Nap* (New York: Simon & Schuster, 1986), 243.

p86 *filmmaker Laura Sky* Laura Sky, *Working Lean: Challenging Work Restructuring* (Toronto: Skyworks for the Canadian Auto Workers, 1990).

p87 *"instrumental action"* Nigel Thrift, "Owners' Time and Own Time: The Making of a Capitalist Time Consciousness, 1300–1880," in *Space and Time in Geography* (Lund, Sweden: CWK Gleerup, 1981), 64.

p91 *half-life of an innovation* Nigel Thrift, "Performing Cultures in the New Economy," *Annals of American Geographers,* 90:4 (2000), 676.

p92 *"non-stop shock treatment"* Mark Dery, *Escape Velocity: Cyberculture at the End of the Century* (New York: Grove Press, 1996), 296.

p93 *"carpet bomb"* Danylo Hawaleshka, "Sick and So Very Tired," *Maclean's,* 15 April 2002, 44.

p94 *"general cognitive impairment"* U. Vollmer-Conna et al., "Cognitive Deficits in Patients Suffering from Chronic Fatigue, Acute Infective Illness or Depression," *British Journal of Psychiatry* (October 1997), 380.

p94 *"depressing period"* Hawaleshka, 44.

p97 *Bowling Alone* Robert D. Putnam, *Bowling Alone: The Collapse and Revival of American Community* (New York: Simon & Schuster, 2000).

CHAPTER 4: VIRTUAL WORLDS AND DESERTING THE REAL

There's a wealth of material on virtual reality and speculation about its social and cultural implications, and some of my favourite authors and titles include Scott Bukatman, *Terminal Identity: The Virtual Subject in Postmodern Science*

Fiction (Durham: Duke University Press, 1993); Donna J. Haraway, "Cyborgs and Symbionts" in *The Cyborg Handbook,* Chris Hables Gray, ed. (New York: Routledge, 1995); Rosanne Allucquere Stone in *Cyberspace: First Steps,* Michael L. Benedikt, ed. (Cambridge, MA: MIT Press, 1992); more technically, Tolga Capin et al., *Avatars in Networked Virtual Environments* (New York: John Wiley & Sons, 1999); John Vince and Rae Earnshaw, *Virtual Worlds on the Internet* (Los Alamitos, CA: IEE Computer Society, 1998) and, in a geopolitical context, Paul N. Edwards, *The Closed World: Computers and the Politics of Discourse in Cold War America* (Cambridge, MA: MIT Press, 1996).

p102 *"post-human"* Stelarc, "From Psycho-Body to Cyber-Systems: Images as Post-human Entities," *Virtual Futures: Cyberotics, Technology and Post-Human Pragmatism,* Joan Broadhurst Dixon et al. (London: Routledge, 1998), 123.

p105 *interactive computer games* Stephen Kline et al. *Digital Play: The Interaction of Technology, Culture, and Marketing* (Montreal: McGill–Queen's University Press, 2003), 12.

p105 *"in game"* Clive Thompson, "Game Theories," *Walrus,* June 2004, 42.

p105 *advertising spots* Kline et al., 20.

p106 *e-stores* Kline et al., 284.

p106 *pumping their stationary wheels* Ken MacQueen, "Pedalling Your Way to a High Score," *Maclean's,* 26 January 2004, 45.

p106 *"gross national product"* Thompson, 41.

p106 *production of signs* David Harvey, *The Condition of Postmodernity* (Cambridge, MA: Blackwell, 1989), 287.

p107 *"iron cleats on his shoes"* Wolfgang Schivelbusch, *The Railway Journey: The Industrialization of Time and Space in the 19th Century* (Berkeley: University of California Press, 1986), 52.

p108 *sixty-yard intervals* Clark Blaise, *Time Lord* (Toronto: Vintage Canada, 2001), 219.

p108 *"synesthetic perceptions"* Schivelbusch, 55.

p109 *"moved him through the world"* Schivelbusch, 64. Writing of the train, he continues: "That machine and the motion it created became integrated into [the traveller's] visual perception.... This vision no longer experienced evanescence: evanescent reality had become the new reality."

p109 *"trivialization of perception"* Schivelbusch, 67.

p109 *"dispersal of attention"* Schivelbusch, 69.

p111 *on to the next thing* Bill McKibben, *The Age of Missing Information* (New York: Random House, 1992), 214.

p112 *hypermedia environment* Ronald Deibert, *Parchment, Printing, and Hyper-media: Communication in World Order Transformation* (New York: Columbia University Press, 1997), 179.

p112 *not just a disembodied* Robert Romanyshyn, "The Dream Body in Cyberspace," *Psychological Perspectives,* Issue 29 (1994), 92.

p112 *"de-centred"* Deibert, 181.

p112 *scathing critique* Somer Brodribb, *Nothing Mat(t)ers: A Feminist Critique of Postmodernism* (Melbourne, Australia: Spinifex Press, 1992).

p114 *"virtual class"* Arthur Kroker and Michael A. Weinstein, *Data Trash: The Theory of the Virtual Class* (Montreal: New World Perspectives, 1994), 108.

p114 *leviathan-like power* Kroker and Weinstein, 108.

p115 *"deeply resentful"* Marshall McLuhan, "Living at the Speed of Light," *Maclean's,* 7 January 1980, 32.

p116 *adapted to this medium* In Canada, First Peoples have created their own television organizations, including the Inuit Broadcasting Corporation and the Aboriginal People's Television Network.

CHAPTER 5: NURSES AND HEALTH CARE

p120 *"kingdom of the sick"* Susan Sontag, *Illness as Metaphor* (New York: Farrar, Straus and Giroux, 1978), 3.

p121 *"triadic" dialogue* Anne H. Bishop and John R. Scudder Jr., *The Practical, Moral and Personal Sense of Nursing* (New York: State University of New York Press, 1990), 172.

p122 *skills and ethics* Joy L. Johnson and Pamela A. Ratner, "The Nature of the Knowledge Used in Nursing Practice," in *Nursing Praxis: Knowledge and Action,* Sally E. Thorne and Virginia E. Hayes, eds. (London: Sage, 1997), 7.

p122 *objective knowledge* Johnson and Ratner, 8.

p122 *"virtuous circle"* Mieke Koehoorn et al., *Creating High-Quality Health Care Workplaces,* CPRN Discussion Paper No. W/14 (Ottawa: Canadian Policy Research Network, January 2002), 1–2.

p122 *healing shrine* Barbara Walker, *The Women's Encyclopedia of Myths and Secrets* (San Francisco: Harper & Row, 1983), 420.

p122 *goddess Rhea* Walker, 766.

p122 *daughters of Asclepius* Jean Watson, *The Philosophy and Science of Caring* (Boston: Little Brown, 1979), 13.

p123 *"the Good Samaritan"* Thelma Pelley, *Nursing: Its History, Trends, Philosophy, Ethics and Ethos* (Philadelphia: W.B. Saunders Co., 1964), 8.

p123 *carnal act* David Cayley, "The Corruption of Christianity: Ivan Illich on Gospel, Church and Society" (*Ideas,* CBC Radio, 2000), 6. Hanif Karim,

a nurse in Vancouver, informed me that an ethic of compassion and
caring is embedded in Muslim sacred texts such as the Koran, and that the
Prophet Muhammad customarily visited the sick. "Thus, visiting the sick
and being in their presence is framed as an act of worship and suggests
God's immanence where there is sickness."

p123 *Roman matrons* Pelley, 9.

p123 *Augustinian sisters* Pelley, 18.

p123 *Ursuline sisters* Pelley, 19.

p123 *Grey Nuns* Vera Campbell, *Elizabeth Bruyere's Great Legacy: Health Care and
Education in Bytown* (Ottawa: The Historical Society of Ottawa, 1988), 8.

p124 *patient histories for interpreting* Barbara Duden, *Disembodying Women:
Perspectives on Pregnancy and the Unborn* (Cambridge, MA: Harvard University Press, 1993), 63–64.

p124 *"to make visible and legitimate"* Geoffrey C. Bowker and Susan L. Star,
Sorting Things Out: Classification and its Consequences (Cambridge, MA: MIT Press, 1999), 29.

p124 *"moral enterprise"* Bishop and Scudder, 32.

p124–25 *"attentive gaze and heartfelt listening"* Thorne and Hayes in Johnson
and Ratner, 37.

p125 *"being all there"* Ginette Lemire Rodger, *A "Z" Nurse-Client Interaction and
its Effects on Stress Level of Clients* (Edmonton: University of Alberta doctor
of philosophy thesis, 1995), 20.

p125 *"are inspired to want to care"* Lemire Rodger, 37.

p125 *"with a very sick population"* Lemire Rodger, 142.

p125 *agents of their own health* Samantha Shatzky, "Cancer Therapy British
Columbia–style," *Ottawa Citizen*, 28 June 2003, A16.

p126 *McDonnell Douglas* Heather Menzies, *Fastforward and Out of Control:
How Technology is Transforming Your Life* (Toronto: Macmillan of Canada, 1989), 83.

p128 *"workload indexes"* Marie L. Campbell, *Information Systems and Management of Hospital Nursing: A Study in Social Organization of Knowledge*
(Toronto: University of Toronto doctor of philosophy thesis, 1984), 71. For
more accessible versions of Campbell's analysis, see "Knowledge, Gendered
Subjectivity and Re-structuring of Health Care: The Case of the Disappearing Nurse," in *Restructuring Caring Labour: Discourse, State Practice and
Everyday Life*, S. Neysmith, ed. (Don Mills, ON: Oxford, 2000) and "The
Structure of Stress in Nurses' Work," in *Sociology of Health, Illness and
Health Care*, 2nd edition, B. Singh Bolaria and H. Dickenson, eds.
(Toronto: W.B. Saunders, 1994), 592–608.

p129 *"care time"* Campbell, 71.

270

p129 *"antithesis of patient-centred care"* Doris Grinspun, "Putting Patients First: The Role of Nursing Care," *Hospital Quarterly* (Summer 2000), 23.

p129 *stress-related ailments* Virginia Galt, "Health Workers Stressed Out, Study Says," *Globe and Mail,* 23 January 2002, A5.

p129 *job satisfaction has* Brian Bergman, "Strains on the Front Lines," *Maclean's,* 8 January 2001, 29.

p130 *dissatisfied with their jobs* André Picard, "Nurses Cite Work Conditions in Quitting," *Globe and Mail,* 7 May 2001, A6.

p132 *withdraw emotionally* Doris Grinspun, "A Flexible Nursing Workforce: Realities and Fallouts," *Hospital Quarterly,* Fall 2002, 79–84.

p132 *passing off shifts* Grinspun, "Flexible." See also Doris Grinspun, "Casual and Part-time Work in Nursing: Perils of Health Care Re-structuring," *International Journal of Sociology and Social Policy,* 2003, 54–70.

p132 *working part-time* Grinspun, "Flexible."

p132 *"psychological contract"* Grinspun, "Flexible."

p133 *effective teamwork* K. Grumbach & T. Bodenheimer, "Can Health-care Teams Improve Primary Care Practice?" *Journal of the American Medical Association,* March 2004, 291(10), 1246–51.

p133 *continuity of caregiver enhances* Grinspun, "Flexible."

p133 *translate for the patient* E. Balka, K. Messup and P. Armstrong, "Indicators of What? Constructing Health-outcome Indicators for a Sustainable Health Care System," (working paper, School of Communications, Simon Fraser University, Burnaby, B.C.).

p134 *patients who died while in Canadian hospitals* Anne McIlroy and Rod Mickleburgh, "Hospital Errors Kill Thousands in Canada, Study Estimates," *Globe and Mail,* 24 May 2004, A11.

p134 *"you have to be pretty sick"* Joel James Shuman et al., *Heal Thyself: Spirituality, Medicine, and the Distortion of Christianity* (Toronto: Oxford University Press, 2002), 97.

p135 *"negative statistic"* Janice Gross Stein, *The Cult of Efficiency* (Toronto: Anansi Press, 2001), 2.

p136 *"rights revolution"* Stein, 60.

p137 *"imposed or negotiated"* Stein, 81.

p137 *"collection and analysis of data"* Roy J. Romanow, *Building on Values: The Future of Health Care in Canada* (Ottawa, Government of Canada, 2002), 54.

p138 *"data gaming," or "data massaging"* Personal conversation with Ellen Balka vis-à-vis research she and colleagues have been doing on statistical health "indicators." The terms refer to what some researchers worry is a growing practice of playing with statistics and data to make them look good. Vancouver, B.C., March 2004.

p138 *"lie sheets"* Balka, Messup and Armstrong, 11.

p139 *narrative and dialogue* I offer an example of this in Chapter 11 when
    I discuss the most-significant change story-based evaluation technique.

CHAPTER 6: MINDING THE COMMON WELFARE

Two works on social work were helpful as background material for this
chapter: Steven Hick, ed. *Social Work in Canada: An Introduction* (Toronto:
Thompson Education Publishing, 2001) and Colleen Lundy, "Historical
Developments in Social Work" in *Social Work and Social Justice: A Structural
Approach to Practice* (Peterborough: Broadview Press, 2004), 19–47.

p141 *the coroner's jury* L. Vanderkoop et al. *Verdict of the Coroner's Jury.*
    (Toronto: Ministry of the Solicitor General, 11 April 2000.)

p141 *"systems failure"* Stein, 138.

p141 *oriented to the world* Mark Kingwell, *A Civil Tongue: Justice, Dialogue and
    the Politics of Pluralism* (University Park, PA: Pennsylvania State University
    Press, 1995), 155.

p142 "pauper auctions" Brereton Greenhous, "Paupers and Poorhouses:
    The Development of Poor Relief in Early New Brunswick," *Social History:
    A Canadian Review,* Vol. 1, No.1 (April 1965), 106.

p142 *pocket any savings* Greenhous, 107.

p142 *George Grant sensed a chill* George Grant, *English-Speaking Justice*
    (Toronto: Anansi Press, 1985). Although this is a recurring theme in
    Grant's writing he spells it out most explicitly in this book of lectures,
    especially on p70 and to a lesser extent on pp46, 57 and 67.

p142 *"reconstitute the social"* Personal conversation with Gillian Walker, profes-
    sor of social work, Carleton University, Ottawa, March 2002.

p143 *"risks and rewards"* Shauna MacKinnon, *Business Transformation Project,
    Government of Ontario: Analysis of a Public-Private Partnership* (Ottawa:
    Canadian Union of Public Employees, 1999), 4.

p144 *"eligibility engine"* CUPE Research Branch, *Welfare Call Centres: Ontario's
    New Assembly Lines?* (Ottawa: Canadian Union of Public Employees,
    2000), 1.

p148 *physically assaulted* CUPE Research Branch, *Overloaded and Under Fire:
    Municipal Social Services Workers* (Ottawa: Canadian Union of Public
    Employees, 2001), 12.

p152 *working until eleven o'clock* Gillian McCloskey, "Charges Dismissed
    Against CCAS Social Worker," *OASW Newsmagazine: The Journal of the
    Ontario Association of Social Workers,* Vol. 27, No.1 (Spring 2000), 1. As well,
    63 per cent of CUPE's *Overloaded* survey respondents reported that they

went to work early or stayed late to keep up with increased workloads, 53 per cent worked more than 150 minutes of unpaid overtime per week, and 73 per cent received work-related after-hours phone calls at home.

p152 *increase in abuse cases* Gay Abbate, "Parents Charged in Assault of Baby," *Globe and Mail,* 7 December 2002, A18B.

p154 "Renées irresponsibility" Jim Coyle, "There Were No Heroes in Jordan's Life," *Toronto Star,* 5 April 2001, B1.

p154 *85 per cent of their time* Simon Cooper, "Child-welfare Time Lost on Paperwork, Report Says," *Globe and Mail,* 23 February 2002, A1.

p154 *Dutch study* CUPE Research, *Welfare Call Centres,* 7.

p155 *Bill is a Toronto children's aid worker* "Bill" is a pseudonym, used here in the interests of confidentiality.

p156 *Bill's colleague, Rod* "Rod" is a pseudonym, used here in the interests of confidentiality.

CHAPTER 7: CHILDREN'S TIME AND ATTENTION DEFICIT DISORDER

p162 *biochemically communicated* Anne McIlroy, "Suffer the Children," *Globe and Mail,* 21 September 2002, F1.

p162 *where the world dwells in us* This is a variation on three concepts: embodiment, as described by Nigel Thrift in "Steps to an Ecology of Place," in *Geographies of Global Transformation,* D. Massey et al., eds. (Oxford, UK: Blackwell, 1998), 314; "dasein," or "being there," associated with Martin Heidegger, and "habitus," associated with Pierre Bourdieu, both described by Christopher Gosden in *Social Being and Time* (Oxford, UK: Blackwell, 1979), 110–17.

p162 *"outer"* Marshall McLuhan, *Counter Blast* (New York: Harcourt, Brace, 1969), 36.

p163 *a circadian rhythm starts up* Gina Kolata, "Finding Biological Clocks in Fetuses," *Science,* Vol. 230 (22 November 1985), 929.

p163 *"rhythms of knowing"* Jeremy Campbell, *Winston Churchill's Afternoon Nap* (New York: Simon & Schuster, 1986), 214.

p163 *pattern of action-reaction* Beatrice Beebe, "The Kinesic Rhythm of Mother-Infant Interactions," in *Of Speech and Time: Temporal Speech Patterns in Interpersonal Contexts,* A.W. Siegman and Stanley Feldstein, eds. (Mahwah, NJ: Lawrence Erlbaum Associates, Publishers, 1979), 23.

p163 *"kinesic rhythm"* Joseph Jaffe and Samuel W. Anderson, "Communication Rhythms and the Evolution of Language," *Of Speech and Time,* 20.

p163 *"split-second world"* Beebe, 24.

p163 *"co-action"* Beebe, 26.

p163 *mother tongue* Noam Chomsky in Campbell, 214.

p163 *"motherese"* Anne McIlroy, "Behind the Tears, A Computer-like Brain," *Globe and Mail,* 27 March 2004, F5.

p163 *"emotional content"* McIlroy, F5.

p164 *"sequence the sounds in words"* Jane Healy, *Endangered Minds: Why Children Don't Think and What We Can Do About It* (New York: Touchstone Books, 1990), 287.

p164 *emergent sense of self* Campbell, 352.

p164 *"Attention with effort"* Virginia I. Douglas, "Stop, Look and Listen: The Problem of Sustained Attention and Impulse Control in Hyperactive and Normal Children," *Canadian Journal of Behavourial Science,* Vol. 4, No. 4 (October 1972), 279.

p164 *"mother's focussed attention"* Erik Erickson, *Childhood and Society* (New York: W.W. Norton, 1963), 248.

p165 *fathers play* Michael Kesterton, "Social Studies," *Globe and Mail,* 13 June 2003, A14.

p165 *"The direct, calm interest"* Gabor Maté, *Scattered Minds: A New Look at the Origins and Healing of Attention Deficit Disorder* (Toronto: Alfred A. Knopf, Canada, 1999), 124.

p166 *mother rats tend to lick* Paul Taylor, "Why Identical Twins Stop Being Identical," *Globe and Mail,* 20 March 2004, F9.

p167 *U.S. accounts for* Peter Breggin, "Peter Breggin, M.D. Testifies Before U.S. Congressional Subcommittee on Psychiatric Drugging of Children for Behavioural Control," *International Center for the Study of Psychiatry and Psychology Newsletter,* (Spring/Summer 2001), 5.

p167 *Canadian physicians* Maria Cook, "Young and Disorderly," *Ottawa Citizen,* 9 December 2001, C3.

p167 *600 per cent* Cook, C3.

p167 *genetics is fingered* David K. Miller and Kenneth Blum, *Overload: Attention Deficit Disorder and the Addictive Brain* (Kansas City, MO: Andrews and McMeel, 1996).

p167 *"because of his short attention span"* Douglas, 260. The pronoun is appropriate given that upward of 80 and 90 per cent of patients are boys.

p167 *"apparently unable to keep"* Douglas, 275.

p167 *serious acting out* Douglas, 274.

p167 *"self-regulatory disorder"* Cook, C5.

p168 *two months before the Nazis* Maté, *Scattered Minds,* 54.

p168 *eight minutes a day* Jay Griffiths, *A Sideways Look at Time* (New York: Penguin Putnam, 1999), 205.

p168 *40 per cent less time* Griffiths, 205.

p169 *children who are stressed* McIlroy, "Suffer the Children," F8. See also Sonia Lupien et al., "Child's stress hormone levels correlate with mother's socioeconomic status and depressive state," *Biological Psychiatry* 48 (2000), 976–80.

p169 *"school ready"* Erin Anderssen, "Trouble Close to Home," *Globe and Mail*, 29 March 2004, A4.

p169 *"feeling for narrative"* Campbell, 387.

p170 *respond to a logo* Eric Schlosser, *Fast Food Nation: The Dark Side of the All-American Meal* (Boston, New York: Houghton Mifflin, 2001), 43.

p171 *watching an average* Healy, 196.

p171 *average of 3.8 hours a week on-line* Media Awareness Network, "Young Canadians in a Wired World, available on-line: www.media-awareness.ca/english/special_initiatives/surveys/index.cfm. See also CBC Health and Science News, "Internet Access Up 36 Per Cent for Tweens—YTV Survey" 30 November 2001, available on-line: www.cbc.ca/story/science/national/2000/10/25/tween001025.html

p172 *"carcooning"* Jessica Johnson, "Carcooning: Life in the Fast-food Lane," *Globe and Mail*, 29 March 2003, L1.

p172 *37 per cent of children* André Picard, "Stress Linked to Obesity in School-age Children," *Globe and Mail*, 2 August 2003, A2.

p172 *don't feel in control* Picard, A2.

p172 *a British study* André Picard, "A New Generation of Couch Potatoes," *Globe and Mail*, 16 January 2004, A3.

p172 *"underworld of children's culture"* Merilyn Simonds quoted in Janice Kennedy, "Why Johnny (and Jenny) Can't Play," *Ottawa Citizen*, 14 June 2003, E3.

p172 *"teach the kids recess"* Kennedy, E5.

p172 *"they don't know how to play"* Gayle Macdonald et al., "School Craze," *Globe and Mail*, 8 February 2003, F2.

p173 *Ursula Franklin* Ursula Franklin, *The Real World of Technology* (Toronto: House of Anansi Press, 1998).

p173 *"This sort of wondering"* Heather Menzies, "The Inexperience of Time," *Ideas*, CBC Radio, 25 October 2000.

p173 *"militarized masculinity"* Stephen Kline et al., *Digital Play: The Interaction of Technology, Culture and Marketing* (Montreal: McGill–Queen's University Press, 2003), Chap. 11, "Designing militarized masculinity."

p174 *high-speed driving simulators* William Illsey Atkinson, "Video Mind Games," *Globe and Mail*, 13 March 2004, F8.

p174 *fighting propensity* Caroline Alphonso, "Do Video Games Breed Violence?" *Globe and Mail*, 18 February 2004, A1.

p174 *link between* Caroline Alphonso, "Schoolyard Bullies Ape Violence on TV, Study for Teachers Finds," *Globe and Mail*, 19 November 2003, A7.

p174 *"cyber bullying"* Alanna Mitchell, "Bullied by the Click of a Mouse," *Globe and Mail*, 24 January 2004, A1.

p175 *"There is no environment"* Richard Lewontin, *Biology as Ideology: The Doctrine of DNA* (Toronto: Anansi Press, 1991), 83.

p175 *signalled danger* Healy, 200.

p176 *"arousal level"* Mark Griffiths and Richard T.A. Wood, "Risk Factors in Adolescence: The Case of Gambling, Videogame Playing and the Internet," *Journal of Gambling Studies*, Vol. 16, No. 2/3 (2000), 212.

p176 *"excitement"* Jeffrey L. Derevensky and Rina Gupta, "Youth Gambling: A Clinical and Research Perspective," *e-Gambling: The Electronic Journal of Gambling Issues*, Issue 2 (2000), 3.

p176 *"jolts per minute"* Morris Wolfe, *Jolts: The TV Wasteland and the Canadian Oasis* (Toronto: Lorimer, 1985), 14.

p176 *"has no outlet"* Healy, 200.

p176 *"the ultimate escape"* Rina Gupta and Jeffrey L. Derevensky, "Adolescents with Gambling Problems: From Research to Treatment," *Journal of Gambling Studies*, Vol. 16, No. 2/3 (2000), 329.

p176 *lots of instant rewards* Cook, C5.

p180 *to be sleep-deprived* Caroline Alphonso, "Sleep Deprivation Leaves Teens Prone to Depression, Study Finds," *Globe and Mail*, 10 February 2004, A7.

p181 *"low-tech, high-touch"* Lillian Katz in Healy, 328.

CHAPTER 8: DRAWING STUDENTS INTO SOCIETY'S CONVERSATIONS

p184 *"implicated participant"* Barbara Adam, *Timewatch: The Social Analysis of Time* (London: Polity Press, 1995), 141.

p184 *read and interpret the world* Judy Hunter, "Implications for Theory" in *Reading Work: Literacies in the New Workplace*, ed. Mary Ellen Belfiore et al., (Mahwah, NJ: Lawrence Erlbaum Associates, Publishers, 2004), 244–45.

p184 *not only reading* David Geoffrey Smith, *Pedagon: Meditations on Pedagogy and Culture* (Bragg Creek, AB: Makyo Press, 1994), 2.

p184 *attunement and attention* Smith, 4.

p184 *a show of hands* See also Caroline Alphonso, "All Work, No Play—and Still Struggling," *Globe and Mail*, 27 March 2004, F7. She cites a study showing an increase in the number of students having to work while also studying at college and university, and she quotes Jim Turk, executive director of the Canadian Association of University Teachers, who says that students are "seriously squeezed."

p185 *related to the labour market* Alphonso, F7.

p185 *"external memory"* Cherry Norton, "Computer Generation Suffers a Memory Crash," *Ottawa Citizen*, 4 February 2001, A1.

p185 *information overload* Norton, A1.

p186 *"wise citizens"* Trudie Richards, professor at Mount St. Vincent University, Halifax, speaking from the floor at the "Conference on on-line education" jointly sponsored by the Canadian Association of University Teachers, the American Association of University Professors, the Canadian Federation of Students and the Quebec University Students' Association, held in Montreal in October 2001.

p186 *"Culture is concerned"* Harold Innis, *The Bias of Communication* (Toronto: University of Toronto Press, 1984), 85.

p186 *retention of information* David Solway, *Lying About the Wolf: Essays in Culture and Education* (Montreal: McGill–Queen's University Press, 1997), 156.

p188 *our ability to* interpret Alan Wolfe, *The Human Difference: Animals, Computers and the Necessity of Social Science* (Berkeley: University of California Press, 1993), 78.

p188 *"relations among things"* Michael Holquist, *Dialogism: Bakhtin and his World* (London: Routledge, 1990), 159. See also M.M. Bakhtin, *The Dialogic Imagination*, trans. C. Emerson and M. Holquist (Austin, TX: University of Texas Press, 1981).

p188 *"I do not use language"* Solway, 163.

p188 *"aphasia"* Solway, 10.

p188 *"chronosectomy"* David Solway, *Education Lost: Reflections on Contemporary Pedagogical Practice* (Toronto: OISE Press, 1989), 78.

p190 *"When language succumbs"* Solway, *Education Lost*, 77. He also links this to time and temporality, adding in an endnote (14) on p. 87: "When time ceases to be *felt* as a psychological constituent of perception and thought, 'communication' becomes sporadic and disorganized."

p192 *more centralized* Janice Newson, "The Decline of Faculty Influence: Confronting the Effects of the Corporate Agenda" in *Fragile Truths: Twenty-five Years of Sociology and Anthropology in Canada*, William Carroll et al. (Ottawa: Carleton University Press, 1992), 230. See also *The Corporate Campus: Commercialization and the Dangers to Canada's Colleges and Universities*, Jim Turk, ed.(Toronto: James Lorimer & Co., 2000).

p193 *academic-time survey* Heather Menzies and Janice Newson, "Academic Time Survey" (unpublished). Some of its findings are reported on in Chap. 2. The full report on the study is in process.

p199 *"banality of evil"* Hannah Arendt, *Eichmann in Jerusalem: A Report on the Banality of Evil* (New York: Viking Press, 1963), 252.

p199 *buzzwords* Arendt, 48. On p. 86 of her text, she links this to Eichmann's propensity for newspeak-like language systems.

p199 *self-knowledge and an inner dialogue* Hannah Arendt, *Responsibility and Judgement* (New York: Schocken Books, 2003), 45.

p199 *Socrates* Arendt, *Responsibility*, 92.

p201 *didn't participate much* Fatemeh Bagherian and Warren Thorngate, "Horses to Water: Student Use of Course Newsgroups," *First Monday* (a peer-reviewed journal on the Internet), Vol. 5, No. 8 (August 2000), 7.

CHAPTER 9: CIVIC DIALOGUE AND NOISY SILENCE

p203 *Seven died* Dennis R. O'Connor, *Report of the Walkerton Inquiry: Part One* (Toronto: Queen's Printer, 2000), 12.

p203 *100 were left with kidney damage* Anthony Reinhart, "The Water May Be Safe, But the Pain Lingers On," *Globe and Mail*, 4 December 2004, A1.

p204 *900 jobs* Bill Tieleman, "Private Profit, Poisoned Water: Walkerton Makes a Case Against Privatization," *National Post*, 5 June 2001, C15.

p204 *minister of health wrote* April Lindgren, "Walkerton Awaits Return of Harris," *Kingston Whig-Standard*, 25 June 2001, 9.

p204 *"information for decision makers"* April Lindgren, "Harris Denies Tories to Blame" *Kingston Whig-Standard*, 30 June 2001, 13.

p204 *"fictionalized"* O'Connor, 7, 197.

p204 *Grade 11* O'Connor, 184.

p204 *Grade 12* O'Connor, 188.

p205 *"is false"* O'Connor, 59.

p205 *"fictitious"* O'Connor, 59.

p205 *effect of pumping* O'Connor, 198.

p206 *"convenience"* O'Connor, 62.

p206 *sampling audit* O'Connor, 301–02.

p206 *fail to pass on* O'Connor, 182. O'Connor went on to say that Stan Koebel "withheld information" and "deceived" both the local health unit and the Ministry of the Environment (p. 183).

p206 *Dr. Kristen Hallett* O'Connor, 67.

p207 *"he thought the water was okay"* O'Connor, 69.

p207 *"We all saw it, the flushing"* "Marcie" and other names used in reporting this interview are pseudonyms. The participants agreed to talk only on the basis of confidentiality.

p209 *"All of us"* Ulrich Beck, *Ecological Enlightenment: Essays on the Politics of the Risk Society* (New Jersey: Humanities Press, 1991), 65.

p209 *mutation of E. coli* Eric Schlosser, *Fast Food Nation: The Dark Side of the All-American Meal* (New York: Houghton Mifflin, 2001), 199.

p209 *"lost sovereignty"* Beck, 66.

p209 *"legitimate totalitarianism"* Beck, 70.

p210 *pseudo form of schizophrenia* Frederic Jameson, "Postmodernism, or the Cultural Logic of Late Capitalism," *New Left Review* (July/August 1984), 71–72.

p210 *"very real chasm"* Thomas Homer-Dixon, *The Ingenuity Gap: Facing the Economic, Environmental, and Other Challenges of an Increasingly Complex and Unpredictable Future* (Toronto: Alfred A. Knopf, 2000), 2.

p210 *key to plastic words* Uwe Pörksen, *Plastic Words: The Tyranny of Modular Language,* Jutta Mason and David Cayley, trans. (University Park, PA: Pennsylvania State University Press, 1995), 76.

p210 *so stretched for time* Tom Barrett, *Walking the Tight Rope: Balancing Family and Professional Life* (Vienna, VA: Business/Life Management Inc., 1994). Some related points about the near dumbing down of political leaders' speech under the stress of crisis are to be found in "Political Rhethoric of Leaders Under Stress in the Gulf Crisis," Michael Wallace et al., *Journal of Conflict Resolution,* Vol. 37, No. 1 (March 1993), 94–107. On the lack of time for reflection and to compose moving speeches, see Rebecca Caldwell, "Where Have All the Orators Gone?" *Globe and Mail,* 23 January 2002, R1.

p211 *mix-and-match sound bites* The Center for Media and Public Affairs, "The Incredible Shrinking Sound Bite," 28 September 2000, available on-line: www.cmpa.com/election2004/index.htm. The CMPA reports that the average sound bite is down to just 7 seconds.

p211 *withering of public trust* Kim Lunman, "Canadians Less Trusting Now, Poll Finds," *Globe and Mail,* 25 June 2003, A8B.

p211 *"social capital"* Robert D. Putnam, *Bowling Alone: The Collapse and Revival of American Community* (New York: Simon & Schuster, 2000), 23.

p212 *"sociological WD-40"* Putnam, 23.

p212 *"rare gift of empathy"* Mark Kingwell, *A Civil Tongue: Justice, Dialogue and the Politics of Pluralism* (University Park, PA: Pennsylvania State University Press, 1995), 120.

p212 *"trained receptivity to the otherness"* Kingwell, 227.

p213 *"very hard most of the time"* Putnam, 189.

p213 *long commutes* Putnam, 212.

p213 *make a difference* Putnam, 284.

p213 *"social capital point of view"* Putnam, 205.

p213 *traditional forms* Putnam, 283.

p213 *"proxy"* Putnam, 156.

p213 *direct-mail* Putnam, 160

p214 *"symbolic affiliation"* Putnam, 154.

p214 *functional networks* Barry Wellman, "The Four Socials of the Internet: Linkages, Capital, Inclusion and Cohesion," *Globalization, Governance and Social Policy: An Expert Roundtable* (Ottawa: Human Resources Development Canada, 2003), 13.

p214 *"fragments of selves"* Wellman, 17.

p214 *"multiple, limited commitments"* Wellman, 64.

p214 *"ties are only weakly connected"* Wellman, 12.

p214 *"extensive, deep, robust"* Putnam, 177.

p218 *"embodied that"* Nelson Adkins, "Introduction to Thomas Paine," *Common Sense and Other Political Writings* (New York: Liberal Arts Press, 1953), xxiii.

p218 *shared time* Jane Jacobs, *Death and Life of Great American Cities* (New York: Vintage Books, 1961).

p219 *"Disengagement and disempowerment"* Putnam, 344.

p219 *Carson predicted* Rachel Carson, *Silent Spring* (New York: Fawcett Crest, 1985).

CHAPTER 10: TAKE YOUR TIME

p227 *tea ceremony* C.A. Simpkins and A. Simpkins, "Cha-no-yu: The Tea Way to Overcome Stress," *Simple Zen: A Guide to Living Moment by Moment* (Boston: Tuttle Publishing, 1999), 101.

p228 *"less wired"* Susan Pinker, "The Silent Treatment," *Globe and Mail,* 1 October 2002, R5.

p228 *"sonic therapy"* Egle Procuta, "Relax, It's Just Music," *Globe and Mail,* 7 November 2000, R6.

p229 *my whole being reaches out* Consciously or not, I'm relating to the ancient Greeks' understanding of optics, in which the eye was understood as a haptic organ, literally reaching out with probing light rays to touch and be touched by what a person saw. See Ivan Illich, *Guarding the Eye in the Age of Show* (University Park, PA: Pennsylvania State University, Science, Technology and Society Studies, Working Papers No. 4, 1994).

p229 *"intimations of deprival"* George Grant, *Technology and Empire: Perspectives on North America* (Toronto: Anansi, 1969), 139.

p229 *"cadence"* Dennis Lee, "Cadence," *Nightwatch* (Toronto: McClelland & Stewart, 1996), 187. He describes cadence as "more flex than content."

p229 *"my body's mind"* Lee, 187.

p231 *"They say that truth moves"* Susan Musgrave "On the Queen Charlotte Islands, Nobody Worries about Manaña," *Globe and Mail,* 28 April 2001, T3.

p232 *"dare to do one thing"* David McFarlane, "Dare to be Different: Do One Thing at a Time," *Globe and Mail,* 4 November 2002, R1.

p234 *attuned to the time of the forest* Judith Stamp in conversation with me in "The Inexperience of Time," *Ideas,* CBC Radio, 25 October 2000.

p235 *"It's about the seams"* Naomi Klein, "Will the Social Fabric Tear?" *Globe and Mail,* 18 February 2002, A15.

p237 *"It's mostly a sense of priority"* Ursula Franklin quoted in "The Inexperience of Time."

CHAPTER 11: TIME FOR DIALOGUE AMD DEMOCRACY

p240 *green world* Northrop Frye, *A Natural Perspective: The Development of Shakespearean Comedy and Romance* (New York: Columbia University Press, 1965), 143.

p241 "The message underlying any medium" Marshall McLuhan, *Understanding Media, Extensions of Man* (Cambridge, MA: MIT Press, 1999), 8.

p242 *ancient Greeks banned* Langdon Winner, "Citizen Virtues in a Technological Order," *Inquiry* 35 (1992), 344.

p243 *as they're paid off and start to run down* Stewart Brand, *How Buildings Learn: What Happens after They're Built* (New York: Penguin Books, 1995), 24, 28.

p243 *"play of public chat"* Jane Jacobs, *Death and Life of Great American Cities* (New York: Vintage Books/Random House, 1961), 61.

p244 *"meat of the question"* Jacobs, 114.

p244 *"weave webs"* Jacobs, 119.

p245 *body language* C. Brannigan and D. Humphries, "I See What You Mean," *New Scientist,* 22 May 1969, 406. Between 75 and 94 per cent of communication is considered to be non-verbal.

p246 *a return to a culture of values* Donald Johnston, "A Call for a Culture of Values, Not Just Rules—from the Corner Office to the Boardroom," *Policy Options/Options Politiques* (Montreal: Institute for Research on Public Policy, November 2003), 26.

p246 *"most significant change"* Jessica Dart and Rick Davies, "A Dialogical, Story-Based Evaluation Tool: The Most Significant Change Technique," *American Journal of Evaluation,* Vol. 24, No. 2 (2003), 137–55.

p247 *"adjust the direction of its attention"* Dart and Davies, 138.

p247 *"evolutionary epistemology"* Dart and Davies, 147.

p248 *"Karoshi widows... right to exist"* Edith Terry, "The Workplace as Killing Field," *The Globe and Mail* (Sept. 21, 1991), D2. The Ministry of Labor has largely foiled this attempt by insisting that "karoshi" claims involve accidents or drastic increases over normal working hours. A 1994 case involving a 42-year-old trucker dying of a heart attack was disallowed because this man's "normal" working hours had averaged around 6,000 hours

(3 times a 40-hour week) for the previous six or seven years. The United National Human Rights Commission has accepted submissions that could result in a formal human-rights investigation on the subject, but little has come of this to date.

p248 *"psychiatric damage"* CUPE Research, *Overloaded and Under Fire: Report of Ontario Social Services Work Environment Survey* (Ottawa: Canadian Union of Public Employees, 1999), 8.

p248 *"Department of Foreign Affairs"* "Bulk Water Removal and International Trade Considerations," *An Act to Amend the International Boundary Waters Treaty Act* (Ottawa: Dept. of Foreign Affairs and International Trade, November 1999), 1.

p248 *Kyoto Protocol* For more information, see www.unfccc.int/resource/ convkp.html

p249 *"performance codes"* Jane Jacobs, *Dark Age Ahead* (Toronto: Random House of Canada, 2004), 154.

p249 *A study of 220 school boards* Jack Mintz, "A Textbook Case of Success," *Globe and Mail,* 6 February 2004, A11.

p250 *East Harlem schools* Mintz, A11.

p251 *We are the experts* Inspired by "We Are the People We've Been Waiting For," the campaign song written by Dennis Schaefer for Saskatchewan Rivers NDP candidate Lon Borgerson in the 2003 Saskatchewan provincial election.

EPILOGUE

p253 *"McWorld"* Benjamin Barber, *Jihad vs. McWorld* (New York: Times Books, 1995).

p254 *peremptorily sacrificed* Estanislao Oziewicz, "War on Terror is Hell on Rights, Amnesty Says," *Globe and Mail,* 26 May 2004, A15.

p254 *surveillance techniques* Ronald Deibert, "The Internet: Collateral Damage?" *Globe and Mail,* 1 January 2003, A11.

p255 *"common sense has lost"* Hannah Arendt, *The Origins of Totalitarianism* (New York: Meridian Books, 1958), 352.

p256 *its will and its pace* Arendt, 467, 466.

p257 *Ninety-nine per cent* Dennis Cauchon, "For Many on Sept. 11, Survival Was No Accident," *USA Today* on-line edition, 20 December 2001, 1. www.usatoday.com.

p257 *"People were fainting"* David Maraniss, "Another Workday Becomes a Surreal Plane of Terror," *Washington Post, Star Tribune,* 21 September 2001, available on-line: www.startribune.com/stories/1576/703150.html

p258 *Kids4Peace* Franke James, "Give Peace a Camp," *Globe and Mail,* 21 August 2004, F7.

INDEX

*H*EATHER MENZIES is the award-winning author of seven books, including the best-sellers, *Whose Brave New World?* and *Fastforward and Out of Control*. She is an adjunct professor of Canadian Studies and Women's Studies at Carleton University as well as a mother, a gardener, a peace and social-justice activist. She lives in Ottawa, Ontario, Canada.